MEN OF ICE

For Jill

MEN OF ICE

LEIF MILLS

The lives of Alistair Forbes Mackay (1878 - 1914)
and Cecil Henry Meares (1877 - 1937)

CAEDMON OF WHITBY

PUBLISHERS

Published by:
Caedmon of Whitby
128 Upgang Lane,
Whitby North Yorkshire
YO21 3JJ

© Leif Mills 2008

ISBN 978 0 905355 69 6

Typeset by:
Croft Publications
The Croft
8 St James Meadow
Boroughbridge YO51 9NW
www.croftpublications.co.uk

Printed and bound by
Smith Settle
Gateway Drive
Yeadon LS19 7XY

CONTENTS

PART I – MACKAY 1878 - 1914   1

    List of Illustrations   viii

    Foreword   ix

    Preface   x

    Introduction   3

    Background   4

    To The Antarctic   7

    South Magnetic Pole   21

    South and North   59

    Canadian Arctic   81

    Aftermath – Notes: Crew etc.   101

PART II – MEARES 1877 - 1937   109

    Introduction   111

    Background   113

    Eastern Adventure   115

    Terra Nova   131

    Army, Navy and Air Force   167

    Japan and Canada   177

NOTES

    Reference Notes   190

    Terra Nova Complement   191

    Bibliography   192

    Glossary   194

ILLUSTRATIONS

Unless otherwise indicated the copyright of all these pictures rests with the exemplary S.P.R.I. of Cambridge

Nimrod	1
Alister Forbes Mackay on the British Antarctic Expedition.	8
Mount Erebus in eruption	18
Carsten Borchgrevink © Caedmon	24
Mackay, David, Mawson at the South Magnetic Pole.	28
Wild, Shackleton, Adams, Marshall after their return from furthest South © Cantebury Museum	70
Mackay with four of the ponies brought from Siberia	84
Pencil Sketch by George Marston of Mackay and Wild playing chess at Cape Royds 1908 © R. G. S.	94
Terra Nova held up in the pack. Dec 1910	109
Meares and Dimitri ready to start South. Nov 1911	110
Meares and Brooke with a Wassu Chief © Caedmon	124
Meares making harness for some of the dogs he brought from Siberia	147
Dimitri and Meares at the blubber stove in the stable, preparing dog pemmican out of seal meat	150
Scott's Hut Plan. Officers and other ranks carefully separated	153
The Tenements, L to R, Cherry-Garard, Bowers, Oates, Meares, (top), Atkinson	153
Meares at the pianola, which also served as a bookshelf and a safe place for delicate instruments	156
Meares upon return from the Beardmore Glacier. Jan, 1912	164
The British Aviation Mission to Japan, 1921 © Medals Research Society	180

Front cover top: Mackay, David, Mawson at the South Magnetic Pole

Back cover: Taylor, Debenham, Gran, Forde

FOREWORD by Sir Ranulph Fiennes

It is now one hundred years since the days of what has been called 'the heroic age of polar exploration' : the age when the North-West Passage was first navigated, two Americans claimed to have reached the North Pole, the South Pole was reached by Roald Amundsen and then by Captain Scott and an attempt was made to traverse the continent of Antarctica.

It is not easy now for people to realise just how difficult and dangerous those explorations were at that time. Much of the area over which they travelled was unknown and had not been travelled before or even seen by man. There were few reliable maps - and none at all in some cases. There was insufficient knowledge of the types of food needed in extreme conditions and little understanding of the dangers of scurvy. Above all there was no means of contact between the explorers and the outside world. There was no radio, no global positioning satellite, no aircraft to lift explorers out of dangerous conditions.

When Shackleton's party was returning from having reached to within ninety seven geographical miles of the South Pole no one at his base at Cape Royds could possibly know where he was at any time or whether anything had happened to them. As it was, they were dangerously late in getting back.

Scott's party was hit by appallingly bad weather on their return journey from the South Pole. No one knew what had happened to them until eight months after they had died when the search party found their bodies. No one knew that where they pitched their final camp was only eleven miles from the supplies of One Ton Depot.

Leif Mills, whose biography of the Antarctic explorer Frank Wild was published nine years ago, has written in this book the lives of two of those explorers from the 'heroic age'. Both Alister Forbes Mackay and Cecil Henry Meares were key members of the expeditions which they joined, yet their names are hardly remembered today and little is known about them. Neither had children and neither wrote any memoirs Both, though, deserve to be remembered for what they did do and the part they played in polar exploration

Mackay and Meares were of a similar age and both had served in the Second Boer War. Mackay served in two polar expeditions but went on to die a tragic death at an early age. Meares served with Scott's second expedition and then served with distinction throughout the First World War. He died at the age of sixty.

I hope this book will go someway to explaining what these two men did and why they did it.

RAN FIENNES

PREFACE

When Mrs Cordelia Stamp of Caedmon of Whitby published my biography of Frank Wild in 1999 I thought that I would not undertake the task of writing another book about polar exploration.

The passage of time – and the coming centenary of the major explorations of Scott, Shackleton and Amundsen – changed that thought.

I started researching the life of Cecil Henry Meares five years ago but came to a stop. Apart from articles in specialist magazines about his travels in China and Tibet, Meares wrote no book, no memoirs nor had he kept a diary. It was difficult to find out much about him. He had no children and I could not track down – in spite of some time spent trying – any of the people to whom his wife, Annie Christina, left legacies in her will. However, with the help of many people and institutions, I was able to pin together the various aspects of his life.

This following biography of Meares is the result. It is not as complete as I would like but it does contain as much information about him as I believe is practically possible to ascertain.

I had similar trouble with trying to unearth some of the details of the life of Alister Forbes Mackay but, again with the help of many people and institutions, I have managed to elicit the key points in Mackay's life. I was helped, of course, by the diary he kept of the journey to the South Magnetic Pole and the diary of the closing stages of the *Nimrod* expedition up to when it left Antarctica.

Again, though, Mackay had no children (he was never married) and apart from his Antarctic diaries left no memoirs or articles about his life. Again, too, attempts at tracking down any relatives also proved negative.

I did, though, manage to solve one particular mystery. Mackay's first name has been variously spelt in polar books, articles and libraries which mention him or have papers relating to him. However, in a hand-written letter from his mother to William Laird McKinlay I noticed her spelling of his name as 'Alister' : I have, of course, therefore used that spelling throughout.

If I have quoted or referred to private correspondence which should have remained confidential, then I am sorry : but all attempts to contact friends or relatives of either Meares or Mackay have come to naught.

Frederike Verspoor, the archivist of the Royal BC Museum in Victoria, British Columbia, has been very helpful to me over the last five years and I am grateful to him for his patience in dealing with my various requests and questions over such a long period.

I am also grateful for the assistance given to me by Derek Oliver of the National Library of Scotland; Andrew Martin of the National Museums Scotland; Sarah Strong, Alasdair Macleod and Joy Wheeler of the Royal Geographical Society:

Naomi Boneham and Lucy Martin of the Scott Polar Research Institute; Peter Helmore of the Orders and Medals Research Society; Paul Carter of the British Library; Dr Philip Towle of the Centre of International Studies in Cambridge; John Fisher, the author of 'Gentlemen Spies' and my former tutor from Oxford, Professor MRD Foot. I also thank the Russo-Japanese War Society and the RAF Air Force Museum at Hendon for their helpful information. Special thanks to John Taylor for his ultra competent cartographic studies fore and aft. Particularly I thank Sir Ranulph Fiennes for agreeing to write a foreword to this book.

Finally, as with my biography of Frank Wild, I wish to thank Cordelia Stamp for all her help, interest and enthusiasm without which this book would not have been possible.

PART I
MACKAY
1878 - 1914

S. Y. Nimrod

INTRODUCTION

At eight o'clock in the morning of Thursday the fifth of February 1914, two Scotsmen, Doctor Alister Forbes Mackay and James Murray, with two companions, left what they had all called 'Shipwreck Camp' in the Arctic Ocean – the place on the ice where the crew of the Canadian ship *Karluk* had put all their equipment, food, clothing, coal and sledges, just before the ship had sunk twenty-six days earlier.

Mackay and Murray had reckoned that they would stand a better chance of reaching land by going themselves (with the two others who had asked to go along with them) and not waiting for the rest of the expedition party of some twenty-two men, one woman, two children, sixteen dogs and one cat. They aimed first for Herald Island, a tiny island some one hundred miles north of the coast of Siberia and then changed their route to Wrangel Island, about forty miles to the west.

Mackay – and Murray – had served on Ernest Shackleton's *Nimrod* expedition of 1907-09. In that expedition Shackleton and three companions had got to within ninety-seven geographical miles of the South Pole – the nearest that anyone had then got to from either the North or South Poles. Mackay and two companions had, in the same expedition, reached the South Magnetic Pole in a harrowing journey in which they had covered 1,260 miles in one hundred and twenty-two days. But experience in the Antarctic, however impressive, did not prepare him or Murray for the awful conditions in the shifting, crashing ice and freezing waters of the Arctic.

Ten days after they had left Shipwreck Camp Mackay's party had met one of the other parties from the camp, which the ship's captain had sent to search for the best way to reach Wrangel Island and lay depots of food and equipment for the main party. Ernest Chafe, the sailor in charge of that party, told the *Karluk*'s captain, Bob Bartlett, when they had returned to Shipwreck Camp, that Mackay's party were all in a pitiable condition and would clearly not make it to Wrangel Island. However much he had pleaded with them , they had refused his offer of joining them to get back to the camp and had insisted on struggling on.

They were never seen again. Mackay was only thirty-five years old.

BACKGROUND

Alister Forbes Mackay was born in February 1878, in Argyllshire, one of seven children of Colonel Forbes Mackay, a regular Army officer with the Gordon Highlanders. He was educated in Edinburgh, then studied biology for a short time in Dundee and then medicine at Edinburgh University.

Mackay always had the reputation of being impatient and hot headed. He was also known as a wanderer. He died tragically at the young age of thirty five but during his life had travelled to South Africa, served there as a trooper and then a policeman, served with the Royal Navy on five different ships and later had been a doctor with a shipping line that sent ships to the Far East and Australasia. He had also been to the Antarctic and the Arctic.

His mother was to write about his life with the Royal Navy and that he must have enjoyed his time with them because, she said of a photograph of him aboard HMS *Research*, "how happy and merry he was in those old days!" But his time in the Antarctic was his finest hour and it must have been a great pity to his mother that his father died before he went with Shackleton.

Mackay never married and there is no record of any serious relationship with a woman. In fact there is not much evidence of strong friendships with anyone. He did keep in touch with a few of those with whom he had been in the Antarctic but, even then, not on a regular basis.

He was physically strong, about six feet tall and well educated. He had a somewhat wry sense of humour and was always forthright in his views. His robust and somewhat cavalier attitude to life must have been one of the reasons that, while still studying medicine at Edinburgh, he volunteered to serve in the City Imperial Volunteers in the Boer War.

When the Boer republics of Transvaal and the Orange Free State went to war with Britain in October 1899 the number of British troops in South Africa were several thousand fewer than the Boers, and they were mainly infantry. Indeed In June of that year the British Army in South Africa was less than ten thousand and the Boer Army was more than three times as big. Even after war broke out the Boers not only had several thousand more men but also had the immediate potential of many more soldiers as all men in the two Boer Republics, between the ages of sixteen and sixty, were liable for conscription into military service; and they were available locally. British troops had to travel six thousand miles to get to South Africa. The other crucial difference was that many of the Boer soldiers had horses and thus were much more mobile than the British regiments of foot.

Although the British Army was to be rapidly expanded, and the number of troops was, by 1901, to become over a quarter of a million – far outnumbering the Boers – significant numbers of those were from Rhodesia, Australia, New Zealand and Canada and the other two South African colonies – Cape Colony and Natal. Eventually the superior British and other Empire numbers were inevitably able to force the Boers to surrender at the end of May, 1902. Yet in the first few months of what was known as the Second Boer War (the first was from 1880 to 1881) the Boers secured several defeats of the British, particularly with three victories in December of 1899 in what was called 'Black Week'.

The public outcry in Britain that followed the military defeats made the Government replace General Sir Redvers Buller as Commander-in-Chief, South Africa by Field Marshal Lord Roberts. It also led to the rapid expansion of yeomanry regiments in Britain – soldiers with horses – as Imperial Yeomanry. Altogether some ten thousand men were enlisted into the Imperial Yeomanry. They were specifically raised for service in South Africa as mounted troops. One of those units was the City Imperial Volunteers (CIV) and Alister Forbes Mackay was one of those early volunteers.

Mackay was still a medical student when he volunteered and joined the CIV in January 1900. He served for most of that year. The CIV was engaged in many victorious battles with the Boers. The first significant victory was at the Battle of Paardenberg (just near the border of Cape Colony with the Orange Free State) at the end of February 1900; the most notable was the Battle of Doonkorp in May 1900 where they successfully dislodged the Boers from a high ridge position when the Gordon Highlanders had failed to do so.

But by September in 1900 there were rumours that the CIV would shortly be sent home. The initial enthusiasm for warfare on horseback waned as the soldiers of the CIV were at most times now engaged in monotonous patrolling.

A number of those who enlisted in the CIV were Members of Parliament, peers, stockbrokers and journalists and they became disillusioned with the policy of burning Boer farms which Roberts had introduced in the summer (but was to stop in November of 1900). Although the recruits had signed on for at least a year, or the duration of the war, significant numbers left by the autumn of 1900 – some to return home and others to join one of the local police forces – and by October only one third of the original numbers were still serving. Then the City Imperial Volunteers as a unit returned home to London at the end of October 1900.

Mackay resigned from the CIV and joined the reorganised South African Constabulary which General Robert Baden Powell commanded. Baden-Powell, the hero of the eight month siege of Mafeking (where his force of twelve hundred men successfully held out against a much superior Boer force) had been tasked by Field Marshal Roberts with the creation of a Constabulary of some six thousand men. They were to control and maintain law and order in those parts of the Transvaal and Orange Free State which the British Army had now secured.

A number of Baden-Powell's Police stayed behind in South Africa after the war and became farmers. Mackay served for a few months and then went back to Edinburgh to complete his medical studies. He graduated from Edinburgh University late in 1901.

He then returned to the battle front in South Africa as a civil surgeon . The Boers finally surrendered in 1902 and signed the peace treaty at Vereeniging on 30 May of that year. Mackay returned home after that.

He had had served in both the CIV and the Baden-Powell Police with distinction and had been awarded the clasps of Cape Colony, Orange Free State and Transvaal to the Queen's South Africa Medal.

During his time with the City Imperial Volunteers he had been struck in the head by a horse. Years later his doctor was to suggest that his excessive drinking was partly due to the effects of that.

Once back in Britain Mackay then joined the Royal Navy and served in five different ships as surgeon – from small gun boats to a 15,000 ton battleship, HMS *Majestic*.

His last naval service was with HMS *Penguin*, a twin screw gun boat, which he joined in October 1906. Then later the following year Shackleton announced his plans for the Antarctic and Mackay was recommended for inclusion on the expedition.

TO THE ANTARCTIC

TO THE ANTARCTIC

Ernest Shackleton had wanted to have his own Antarctic expedition ever since he was sent home – unfairly he thought – by Robert Falcon Scott after little more than one year with Scott's *Discovery* expedition of 1901 to 1904. He received over four hundred applications to join his proposed Antarctic expedition. Most were from newspaper advertisements but some from individual recommendations from amongst those supporting the expedition. He decided on a small party and particularly wanted two of the men to be qualified doctors.

He chose Eric Marshall, public school, Cambridge University and St Barthomolew's Hospital, who had qualified as a surgeon in 1906. Marshall would also be the expedition's cartographer, surveyor and photographer. The other doctor, who also acted as the expedition's second biologist, was Alister Forbes Mackay. The principal biologist was James Murray.

Of the fourteen men who finally formed the shore party only two – apart from Shackleton – had previous Antarctic experience : Frank Wild and Ernest Joyce. They had both been with Scott's *Discovery*. Joyce had sixteen years service with the Royal Navy and Wild had six, but also ten years with the merchant navy. Joyce was to be in charge of dogs, sledges and equipment and Wild in charge of the food and supplies. Wild had been friendly with Shackleton during the *Discovery* expedition and Shackleton had a high opinion of his qualities – an opinion that was more than justified by Wild's actions and experiences both on this expedition and, later, on Shackleton's *Endurance* expedition of 1914 to 1917.

Both Wild and Joyce were sent by Shackleton on a three week course in printing techniques just prior to the start of the expedition (at the offices of Sir Joseph Causton & Sons, Ltd. in Hampshire). The printing firm lent a letter press and other printing equipment to take with them to the Antarctic. Shackleton had the idea of having a book published during their stay in Antarctica – the first book ever to be printed there. It was to be named *Aurora Australis*.

Ten members of the expedition wrote articles for the book, five under their own name and the other five anonymously. The book totalled one hundred and twenty pages and George Marston, the artist of the expedition, drew eleven illustrations for it. Wild and Joyce had brought the printing paper and ink with them. Bernard Day, the expedition's motor mechanic, made book covers from the plywood of the packing cases and seal skin was used on the front and back boards.

One of the ten articles was by Mackay : 'An interview with an Emperor', an amusing short story about Mackay and an unnamed companion going to Cape Crozier – where a breeding ground for Emperor penguins was meeting up with a six foot tall Emperor penguin who spoke in a Scottish accent. Mackay had signed the article with his initials.

Up to one hundred copies of each page were printed (the exact number is not known) but only between twenty-five and thirty copies were sewn and bound. Each member of the shore party had one. It is not known now how many copies survive but one of those copies was sold at auction in London recently for many thousands of pounds.

As ever, Shackleton was short of money for what he called the 'British Antarctic Expedition' and again time was also short. He had to take his second choice of a boat – the *Nimrod*, a forty-one year old wooden sealer, originally built in Dundee, but then based in Newfoundland in Canada. Shackleton paid £5,000 for the ship and she arrived in London in mid-June of 1907. *Nimrod* was small – almost too small – and needed an extensive overhaul.

By the beginning of August preparations for the expedition were as advanced as they could be. Fifteen Manchurian ponies would be shipped from China to New Zealand. The *Nimrod* would carry a motor car – the first to go to Antarctica. It was built by the Arrol Johnstone Motor Car Company, had a twelve to fifteen horse power air cooled engine and was capable of a speed of sixteen miles per hour. Shackleton wanted to try it out and, if possible, use it for taking heavy loads over the Great Ice Barrier. The front pair of wheels were shod with wood and could have a sledge attached to them; the rear wheels were adapted to take spikes to give greater traction on snow and ice.

As it turned out the motor car was of little use. Although it had two petrol tanks and, when full, would carry sufficient fuel for a journey of three hundred miles, and although the idea was also to connect the exhaust from the engine to a snow melter and thereby make water, the car basically failed. It would get bogged down in the snow and often break down.

Mackay sprained his wrist while trying to start it with the crank handle. In fact that was just before starting off on the journey to locate the South Magnetic Pole and Mackay therefore began that journey with his arm in a sling.

The *Nimrod* sailed from London and called at Cowes in the Isle of Wight where King Edward VII and Queen Alexandra came on board to wish the expedition well and present Shackleton with a Union flag to carry to the South Pole. The *Nimrod* then went to Torquay and from there, on the seventh of August, 1907, sailed for New Zealand.

The only two members of the proposed shore party of the expedition to sail on the *Nimrod* down to New Zealand were Mackay and Murray. They were to undertake zoological studies and also do some ocean dredging on the voyage. Frank Wild and seven of the others sailed to New Zealand on the White Star liner *Runic* (a ship that mainly carried emigrants to Australia and New Zealand). Shackleton sailed separately to New Zealand as did the other member of the expedition from Britain, Sir Philip Brocklehurst, the nineteen year old Cambridge student, who was to serve as an assistant geologist and surveyor. Brocklehurst's family had given a substantial sum of money to Shackleton for the expedition.

Before Shackleton had joined the *Nimrod* in New Zealand he had stopped at Australia. In Sydney, Professor Edgeworth David, of Sydney University, geologist and surveyor, had written to Shackleton asking if he could go down to the Antarctic in the *Nimrod* and then return with the ship when it sailed back. Shackleton had agreed, though when the *Nimrod* reached Antarctica he persuaded David to join the shore party. At forty-nine years old David was the oldest in that party.

Through David's good offices Shackleton recruited two further Australians – Douglas Mawson, a lecturer at Adelaide University, as physicist, and Bertram Armytage, a Cambridge graduate and who had fought in the Boer War, who was to be in charge of the ponies.

Fifteen ponies had been bought in China and they came by boat to New Zealand. Because of quarantine restrictions the ponies could not be put ashore on the mainland of New Zealand. Instead they were taken to Quail Island, a small island off Port Lyttleton at the bottom of New Zealand's South Island, which was used as a quarantine station. There Mackay was put in charge of the ponies – presumably because he had some experience of horses during his time in the Boer War, and none of the others did – and was charged with breaking them in and preparing them for pulling sledges in harness. When the *Nimrod* got to Antarctica then Armytage took over responsibility.

The *Nimrod* was a very small ship and with its complement of nine dogs (Shackleton was later to regret not taking more), timber for the hut, sledges, stores, food, the motor car, coal for the ship's engine, Shackleton and the fourteen men of the shore party, as well as the crew, there was little room for the ponies.

Shackleton chose what he thought were the best ten of the fifteen ponies Mackay had been looking after. There were five stalls for the ponies on either side of the ship's main deck. They were narrow and hardly adequate. One pony was so badly chafed and sore from the rubbing and knocking against the stalls that when the *Nimrod* reached Antarctica it had to be shot. Earlier on in the voyage, during a fierce storm that lasted nine days, one of the ponies had slipped over as the ship was rolling and could not get to its feet again. It was so badly injured that Shackleton had ordered it to be shot.

Of the eight ponies that were landed in Antarctica, three died from eating too much sand which lay on the ground where they were picketed. The ponies had insufficient salt given to them and had eaten the sand because of its saline flavour through the wash of sea water on to the beach. The fourth pony then died through eating some wood shavings which had contained some chemicals that caused corrosive poisoning.

That left four ponies and it meant that Shackleton had to abandon one of the three aims for the British Antarctic Expedition. One aim was for a small party with ponies to explore King Edward VII Land, by the eastern end of the Great Barrier, and from there go westward but this would no longer be possible. It also meant that there could only be four men on the Southern Party that would aim to reach the South Pole and not six as Shackleton had originally hoped. His third objective – to send a party (the Northern Party) to locate and reach the South Magnetic Pole – would have to be undertaken by man-hauling : though, at first, Shackleton hoped that the motor car would be able to go with them for some time.

There had also been a tentative proposal to send a small party to Cape Crozier, in the McMurdo Sound, to examine the breeding grounds of Emperor penguins but this was called off because of the lack of sufficient ponies. Later, as part of Scott's *Terra Nova* expedition, Edward Wilson, 'Birdie' Bowers and Apsley Cherry-Garrard were to man-haul a sledge to the penguin breeding ground in the depths of the Antarctic winter. Cherry-Garrard called this journey 'the worst journey in the world' and his book of the same name remains one of the finest books on Antarctic exploration.

Because of the inadequate amount of coal that the *Nimrod* could carry it was clear that the ship would not be able to get to Antarctica and back under her own power, so Shackleton arranged for a steel built steamer ship, the

Koonya, to be hired to tow the *Nimrod* from New Zealand the fifteen hundred miles to the Antarctic Circle – 66°33' south.

Before the *Nimrod* sailed Shackleton had arranged for another 'first' for his expedition. He had agreed with the Prime Minister of New Zealand that he would act as an official postmaster and would stamp letters from his base in Antarctica. In fact when the *Koonya* left the *Nimrod* she took some stamped letters and further letters were taken back by the *Nimrod* when she returned both in 1908 and then on the final voyage back to New Zealand in 1909.

The *Nimrod* left New Zealand on the first of January, 1908. The *Koonya* towed her for the first fifteen days and then turned back. The *Nimrod* came within sight of the Great Ice Barrier on the twenty-third of January. Shackleton had hoped to establish his base on King Edward VII Land, by the western edge of the Barrier, particularly to see if it was possible to land at what was called Balloon Inlet. He had promised Captain Scott in England that he would not use the area by McMurdo Sound where Scott's *Discovery* expedition had established its base – near what Scott had called Hut Point.

That Scott had insisted on such an undertaking does seem extraordinary today but at the time Shackleton had little alternative but to agree. The Royal Geographical Society was already backing Scott's idea for his own second Antarctic expedition and Shackleton had no wish to offend them or appear as though he was exploiting Scott's earlier discoveries.

As it turned out Shackleton felt it was dangerous to land anywhere along the Barrier. One spot, the indentation in the Barrier where Scott had launched the balloon carried on the *Discovery* ('Balloon Inlet') had now disappeared and he could see that the calving of the ice into the sea all along the Barrier could make locating his expedition's hut on it extremely dangerous. He looked at landing at the Bay of Whales on the Barrier but thought this was too dangerous (though Roald Amundsen, the famous Norwegian explorer, was to use it as his base in 1911 to 1912 for his successful South Pole attempt). He then decided he had no alternative but to land by McMurdo Sound. Ice stopped the ship from going straight to Hut Point so she stopped at Cape Royds, some twenty-three miles north of Scott's *Discovery* hut at Hut Point. It was a breach of the undertaking he had given to Scott but Shackleton considered he had had no choice.

While the stores and equipment were being landed from the *Nimrod* an appalling accident happened to the ship's second officer, Aeneas Mackintosh, whom Shackleton was intending to join the shore party. When lifting a packing case from the cargo hold, the hook attached to it slipped and struck Mackintosh in the right eye. The pain was excruciating and Marshall, assisted by Mackay and Dr Mitchell, the ship's doctor, then immediately administered chloroform and took out the eye. It meant that Shackleton had to send Mackintosh back to New Zealand with the *Nimrod.* and could play no further part in the shore party – though he did sail with the *Nimrod* when it returned to Cape Royds in January 1910. Mackintosh was also to return to Antarctica later as leader of the Ross Sea Party of Shackleton's *Endurance* expedition

The *Nimrod* had reached Cape Royds at the end of January and for the next three weeks the shore party and the ship's crew landed all the provisions, stores, timber, stoves, sledges, scientific equipment, ponies and dogs, built shelters for them and erected the hut. The hut was divided up into cubicles for two men each – apart from Shackleton who had his own separate cubicle at one end. Mackay shared his cubicle with William Roberts, the expedition's cook and former chef at the Army & Navy Club in London.

Each of the shore party would try and make their cubicle somewhat different from the others and put up shelving, books, pictures and arrange space for their clothes.

Shackleton described the shore party's amusement at the setting up of the cubicle of Mackay and Roberts [HoA 1 p146] :

"Beyond the stove, facing the pantry, was Mackay's and Roberts' cubicle, the main feature of which was a ponderous shelf, on which rested mostly socks and other light articles, the only thing of weight being our gramophone and records. The bunks were somewhat feeble imitations of those belonging to No. 1 Park Lane [the name given to the cubicle shared by Marshall and James Boyd Adams, the expedition's second in command] and the troubles that the owners went through before finally getting them into working order afforded the rest of the community a good deal of amusement.

"I can see before me now the triumphant face of Mackay, as he called all hands round to see his design. The inhabitants of No. 1 Park Lane pointed out that the bamboo was not a rigid piece of wood, and that when Mackay's weight came on it the middle would bend and the ends would jump off the supports unless secured. Mackay undressed before a critical audience, and he got into his bag and expatiated on the comfort and luxury he was experiencing ."

Shackleton then noted that Mackay's bed crashed "and was half on the ground, one end of it resting at a most uncomfortable angle. Laughter and pointed remarks as to his capacity for making a bed were nothing to him; he tried three times that night to fix it up, but at last he had to give it up as a bad job." Shackleton did add, though, that "in due time he arranged fastenings, and after that he slept in comfort."

During the Antarctic winter, in such a confined space as their small hut, it is not surprising that tempers sometimes flared. In each cubicle two men had to live closely together. Most of them did get on with each other but Mackay had not met Roberts before and the two came from completely different backgrounds. Murray, with whom he had travelled down to Antarctica in the *Nimrod*, shared with Raymond Priestley, the young assistant geologist and there was no one other than Roberts with whom Mackay could share.

In early August, towards the end of the Antarctic winter, Mackay's temper snapped with Roberts. It appeared that the cook had put his feet up on Mackay's sea-chest to lace up his boots. Mackay got his hands round Roberts' neck and made an attempt to strangle him. Mawson intervened and stopped the matter from getting any worse. In Marshall's diary [3Aug1908] was the comment that Shackleton (whom Marshall frequently criticised in his diary) was "in a regular panic about it & threatens he will shoot [Mackay]. This is the 2nd time he has said. He is so easily frightened that he is not to be trusted with a pistol … Mac quite all right but slightly eccentric & quick tempered". There is no record anywhere of the first occasion to which Marshall referred.

By the time the hut had been completed and all the stores and equipment unloaded from the *Nimrod* and before the winter set in , Shackleton was anxious for some other activity that would be popular and appear worthwhile – and, above all, keep the men occupied. The *Nimrod* had sailed back to New Zealand on the twenty second of February and the hut was finally sorted out by the beginning of March. Shackleton then proposed that a party of the men should climb Mount Erebus, some fifteen miles away from their hut.

It was probably David who first suggested this, particularly as he would want to examine the geology of what was still an active volcano, and Shackleton adopted the idea enthusiastically.

Sir James Clark Ross had visited this part of Antarctica in his voyage of 1839 to 1843 with two ships. His expedition – in 1840-1841 – was the first ever to see the Great Ice Barrier (which was later named the Ross Ice barrier after him). He was the first man to visit the open water by the ice Barrier (later named the Ross Sea) and also see the two volcanic mountains which he named, Mount Erebus and Mount Terror after his two ships. He estimated the height of Mount Erebus at 12,367 feet. He also entered the bay, by the volcanoes, called McMurdo Sound – and named the land to its east as Victoria Land.

Shackleton explained his decision [HoA 1 p170] by stating that open water between them and Hut Point prevented them from laying a depot for the proposed Southern Party south of Hut Point before the onset of the Antarctic winter and so "we began to seek some outlet for our energies that would be useful in advancing the cause of science, and the work of the expedition. There was one journey possible, a somewhat difficult undertaking certainly, yet gaining an interest and excitement from that very reason, and this was an attempt to reach the summit of Mount Erebus. For many reasons the accomplishment of this work seemed to be desirable.

"In the first place, the observations of temperature and wind currents at the summit of this great mountain would have an important bearing on the movements of the upper air, a meteorological problem as yet but imperfectly understood. From a geological point of view the mountain ought to reveal some interesting facts, and apart from a scientific considerations, the ascent of a mountain over 13,000 ft. in height, situated so far south, would be a matter of pleasurable excitement both to those who were successful as climbers and to the rest of us who wished for our companions' success".

Shackleton chose David, Mawson and Mackay for the summit party. Then he named Adams, Marshall and Brocklehurst as the support. Curiously he named Adams as in command of the whole party, though after the supporting party went back, he named David as leader. In fact two of the three man support party were also to reach the summit with the original three men selected.

Though the idea of climbing Mount Erebus was a good one it was clear that no thought had gone into this beforehand. None of the six men had any mountaineering experience. None had proper mountain boots nor even sufficient rucksacks. David put strips of seal leather on his soft soled boots; most of the others put nails on their boots to use as crampons. Makeshift rucksacks were made from pieces of rope. David, Mawson and Mackay each had separate sleeping bags. The support party had a three man sleeping bag.

It was a motley party that set off in the morning of the fifth of March 1908. They pulled one sledge with a load of 560 pounds. Shackleton and the remainder of the shore party waved them off.

By the evening the party had marched seven miles and climbed to 2,750 feet. On the evening of their second day, the sixth of March, they had gone three miles but had climbed to an altitude of 5,630 feet. The temperature was dropping and the climb was getting harder. That night the temperature fell to minus 28° Fahrenheit.

On the following morning Adams, although due to turn back soon, decided that the supporting party should go with the others to the summit. They had no rucksacks and some had not put nails in their boots but he thought it would be worth it. Shackleton later claimed that he had left the decision to Adams' discretion. In the event there was nothing he could have done about it.

They decided to leave the sledge there and proceed carrying their own loads on their backs, each weighing about forty pounds. David, Mawson and Mackay each carried their own sleeping bags. The three man sleeping bag of the support party was folded up with ropes and string and carried by each of them in turn. They tried to carry their tent poles with them but quickly abandoned the idea : it was just too difficult and the climb was getting steeper, as much as some thirty-four degrees at times. At night they just stretched the tent cloths over the sleeping bags.

On the evening of their third day, the seventh of March, they had climbed to 8,750 feet. Then a blizzard struck and they could not go any further for the next thirty six hours.

When the blizzard had died down they started again early in the morning of the ninth of March. They were all roped together, partly because of the steepness of the climb and partly because some of the party had no nails in their ski boots to use as crampons. Later that day Mackay offered to go by himself up a steep slope to see if there was a quicker way to the summit. After a while the rest of the party heard him cry out. Mackay had found cutting steps with his ice axe, and carrying his load on his back, too much and he almost fainted. Marshall determined that Mackay was suffering from altitude sickness.

Brocklehurst also suffered from altitude sickness and found that his feet were becoming frost bitten. Marshall inspected his feet when they stopped at lunch time: his two big toes were black and the other toes were also affected. Marshall and Mackay tried to rub them to give him some warmth. Then they gave him some dry socks, put some finnesko boots on his feet and wrapped him in the three man sleeping bag. It was evident that he could not go any higher.

The other five men continued their climb and that evening reached 11,400 feet.

They had passed the old crater near the summit and the next morning, the tenth of March, at ten o'clock, they reached the still active crater at the summit. Marshall determined, by aneroid and hypsometer, that the mountain was 13,370 feet above sea level. The depth of the volcanic crater Mawson estimated to be some nine hundred feet.

They had done it – the first men to climb a mountain in Antarctica.

That same day they started down, picked up Brocklehurst and the three man sleeping bag on the way and made good progress, partly by sliding down some of the snow slopes. That evening they camped where they had left the

sledge and a small depot of food and equipment. The following morning, the eleventh of March, they started at five-thirty in the morning and again through a combination of climbing and sliding down, with many falls on the way, made good progress. By midday they were back at their hut, to the delight of Shackleton and the rest of the shore party. They had come down from the summit in under two days.

Back at the hut, Marshall then amputated one of Brocklehurst's big toes, with Mackay acting as the anaesthetist.

When asked by Shackleton about the medical condition of the men, Marshall made it clear that he did not think Brocklehurst should now go on the Southern Party to try and reach the Pole. He also had doubts about the fitness and stamina of Joyce : although Joyce was later to claim that Shackleton had more or less promised him a place on the Southern Party, it was not to be. In any event there were only four ponies left.

Shackleton then decided to take Adams, Marshall and Frank Wild with him to try for the South Pole.

Shackleton, from the reports of the party that climbed Mount Erebus, thought that David, Mawson and Mackay had done well together and so he named them as the three man party, the Northern Party, to try for the other remaining objective of the expedition – to locate and reach the South Magnetic Pole. David and Mawson, he reckoned, would be able to do a lot of geological and survey work and Mackay was the second doctor on the expedition and was strong and capable.

The South Magnetic Pole is the point on the Earth's surface where the direction of the Earth's magnetic field is vertically upwards. The magnetic dip, the angle between the horizontal plane and the Earth's magnetic field lines is $90°$ at the magnetic poles. The North and South Magnetic Poles are not fixed and move continually. The Norwegian explorer, Roald Amundsen, had been the first to locate and reach the position of the North Magnetic Pole during his voyage – again, the first ever – through the North West Passage from the Atlantic Ocean (via the sea around Greenland and northern Canada) to the Pacific Ocean (via the Bering Strait) from 1903 to 1906. Shackleton's men would be the first to locate and reach the South Magnetic Pole.

As soon as the winter drew to a close Shackleton sent out parties to Hut Point to store food and supplies there and then lay depots further south for the Southern Party, the furthest being one hundred miles south. It was essential that this was done and further depots would also have to be laid as the Southern Party neared the Pole.

Mount Erebus in eruption.

Two small depots were also laid a few miles towards South Victoria Land for the Northern Party. With all that activity it was not until early in October that the Northern Party for the South Magnetic Pole was finally able to set out. David, as leader of the party, had wanted to go earlier but they had not been ready to leave Cape Royds as early as they should. As it turned out their later start did not materially affect their journey or their return. In any event, it would be a close run thing.

SOUTH MAGNETIC POLE

SOUTH MAGNETIC POLE

After an early breakfast on the fifth of October, 1908, the Northern Party of David, Mawson and Mackay started off. Day, Priestley and Roberts – with the motor car – started with them. After two miles it was clear that the motor car was not going to be of much use, particularly with snow falling and thus making it difficult for the car wheels to grip the surface, and David agreed they should go back to Cape Royds. He and his two companions would have to make the journey without any assistance from the motor car; and there were no ponies to go with them.

There were now the three men, with two eleven foot sledges, one tent, one three man sleeping bag, food – mainly plasmon biscuits, pemmican, sugar and chocolate – and clothes, boots and equipment. The total weight they had to pull was some seven hundred and ten pounds. They estimated that the distance between Cape Royds and where they thought the precise point of the South Magnetic Pole would be located was about five hundred miles.

It was Captain James Clark Ross who first charted the coast of what he called South Victoria Land in his expedition of 1839 to 1843 with his two ships HMS *Erebus* and HMS *Terror* and had named the mountains along the coast the 'Admiralty Range'. He also named the headland by the beginning of South Victoria Land as Cape Adare. He estimated that the Magnetic Pole would be somewhere inland in South Victoria Land.

The next expedition to visit the area was over fifty years later when the Norwegian Henrik Bull, in his small wooden steamship the *Antarctic,* sailed there with a small party in 1894 to 1895. Three men of that expedition landed briefly at Cape Adare.

Nearly five years later Carsten Borchgrevinck, another Norwegian explorer, went there as leader of a British expedition from 1898 to 1900 in the ship *Southern Cross.* Borchgrevinck had been a member of Bull's earlier expedition and later claimed to have actually been the first man to have set foot on the Antarctic mainland when the three men landed at Cape Adare. This has been the subject of much dispute (both within Borchgrevinck's party itself and with Antarctic historians) with some claiming that the first landing was made many years earlier by a sailor from a whaling ship on the other side of the continent. This time, however, Borchgrevinck was to claim accurately that his party of ten men were the first people to winter on the mainland. They lived in a hut on Cape Adare from early March 1899 to the end of January 1900 when the *Southern Cross* picked them up.

Louis Bernacchi, an Australian of Italian descent and born in Belgium, was a physicist and magnetician with Borchgrevinck's party and he had estimated the position of the South Magnetic Pole. Although there appeared to have been considerable friction among the ten men wintering party, with particular criticism being made of Borchgrevinck himself, Bernacchi was held to have made a major scientific contribution to the work of the expedition. He was then to join Scott's *Discovery* expedition as physicist and again made estimates of the position of the South Magnetic Pole.

Captain Scott's *Discovery* expedition of 1901 to 1904 was also to visit Cape Adare and the surrounding area. The German South Polar Expedition of 1901 to 1903, led by Professor Erich von Drygalski in the ship *Gauss*, sailed right along the coast of South Victoria Land and beyond to what he named as Kaiser Wilhelm II Land towards where Mawson was later to lead his own Antarctic expedition from 1911 to 1914.

Much of the coast where the three men of Shackleton's expedition were to go, therefore, had been seen and charted, mountains and inlets named and maps drawn but, although Borchgrevinck's men had wintered on Cape Adare, no one had ever travelled on foot along the coast of South Victoria Land and no one had ever been inland from the coast. David, Mawson and Mackay were the first to do so.

Although they had estimated the distance they had to travel as some five hundred miles they were, altogether, to travel one thousand, two hundred and sixty miles in one hundred and twenty-two days – seven hundred and forty miles of which were relay work : pulling one sledge for a distance and then leaving it and going back for the other sledge and pulling that. Until they left one of the sledges and went on with just the one, it meant that for every mile forward they had to do three miles. All the while they were relaying the two sledges they were only averaging four miles onward each day.

They were to travel along the coast and sea ice of the southern edge of South Victoria Land and then turn inland – northwards and upwards for several thousand feet on to the plateau (where the South Magnetic Pole was estimated to be was 7,620 feet high) which stretched far into the distance – when they reached the Drygalski Glacier and head towards where Bernacchi had estimated the South Magnetic Pole to be and where they would try and

fix the position themselves. They knew the Magnetic Pole moved; what they did not know was how far it would have moved since Bernacchi had located it years earlier.

FIENNES AND STROUD

They had no dogs and no support parties. When they turned inland, after reaching the Drygalski Glacier, they would be going where no one else had ever been or even seen. Their record of such a long unsupported mileage was to last for over eighty years. It was not surpassed until Ranulph Fiennes and Mike Stroud in 1992-93 sledged, via the geographical South Pole, from Berkner Island off the Filchner Ice Shelf, by the Weddell Sea, to half way across the Great Ice Barrier. They sledged for one thousand, three hundred and fifty miles unsupported. A great achievement, but they were in radio contact with the outside world and when they could go no further they radioed for assistance and were airlifted out.

David, Mawson and Mackay were on their own with no contact with the outside world. They were due back at Cape Royds by the beginning of February but if they did not make it by then the *Nimrod* would sail along the coast of South Victoria Land and hope to meet them on their way back : if the *Nimrod* failed to see them within a few days then they would again be on their own. The ship would have to go back to Cape Royds because of a limited supply of coal for the ship's engine and also because the ship's captain would, understandably, not wish to spend too long searching for them and thus risk the ship getting frozen in the ice and spending the following winter there. Also the ship had to be back at Cape Royds to pick up Shackleton's Southern Party in time to get away before the ice froze the ship in.

So, if the *Nimrod* did not pick them up, then the outlook would be grim. They would have to try and make it back by themselves to the hut at Cape Royds and hopefully be picked up there. If they were too late and missed the *Nimrod* picking up the Southern Party of Shackleton, Marshall, Adams and Wild, then they would presumably have had to wait for another winter at the hut and hope a ship would fetch them after that ; but that would be a pretty vain hope.

David's account of their journey was published in the second volume of Shackleton's 'The Heart of the Antarctic'. It was a straightforward and quite detailed narrative of their journey. What he did not mention was the friction between the three of them and, in particular, the stress caused by his own behaviour and how, given the nature of the three people, it was remarkable

Carston Borchgrevinck

that they actually got, on the sixteenth of January 1909, to where they thought the South Magnetic Pole was – in Mawson's measurements it was located at latitude 72° 25' south and longitude 155° 16' east – and then made it safely to where the *Nimrod* picked them up on the sixth of February.

Mawson kept a diary (as he was to do on his later Antarctic expeditions) and entered some fairly trenchant comments on David, though he was never malicious in them. Mackay kept a diary from the end of November until the day they were picked up by the *Nimrod*, the sixth of February. He seldom referred directly to problems with David except towards the end of their journey and particularly where he spelt out (in his entry for the third of February) that David would have to hand over the leadership of the party to Mawson or otherwise he would declare him 'physically and mentally unfit'.

They got to the first small depot on their first day and the second depot two days later. On the thirteenth of October they reached Butter Point, an angle in the low ice cliff near the junction of the Ferrar Glacier valley with the main shore line of South Victoria Land, Shackleton had asked them to leave a depot flag at Butter Point stating what progress they had made and when they expected to return there.

They left some clothes and biscuits as a depot and David left two letters – one to Shackleton and one to Priestley. In them David had written that because of slow progress so far, and their late start from Cape Royds, they would not get back to Butter Point before the twelfth of January at the earliest – not the first week of January as originally intended.

After they had left Cape Royds, and before he departed on the southern journey, Shackleton had written out further instructions for Priestley. He, Armytage and Brocklehurst were to go to Butter Point in early December and do some surveying of the area and its geological make-up. Brocklehurst was to take photographs. They would take with them some extra provisions for David and his two companions and they then would wait and meet them on their way back from the South Magnetic Pole – hopefully by the first week in January.

Assuming the two parties did meet up, David was to go back with Mackay and Armytage to Cape Royds leaving Mawson, Brocklehurst and Priestley to carry on with surveying work and, in Mawson's case, to look for what Shackleton called "economic minerals". Priestley was to hand over Shackleton's written instructions to David when they met.

But it was not to be. The timetable was far too tight and David and the others realised fairly early on in their journey that there was no way they could reach the Magnetic Pole and get back to Butter Point by early January, if at all.

David's letters to Shackleton and Priestley were found at the Butter Point depot by Priestley. Priestley, and Armytage did some geological survey work – and Brocklehurst took some photographs – but they found no trace of what Shackleton had termed "economic minerals". They were then later picked up by the *Nimrod* when the ship went from Cape Royds along the coast of Victoria Land to search for the Magnetic Pole party.

From the beginning, Mawson wrote of the difficulties that David caused the other two. The three of them had a large enough load to man haul, the relay work was back breaking and took enormous time, and the three men were each disparate characters. On top of that, a three man sleeping bag, though useful for greater combined body warmth, was not the most comfortable sleeping arrangement. Friction was therefore inevitable. David was twenty years older than Mackay and almost twice as old as Mawson. His deterioration in health during their journey was perhaps bound to happen, given their circumstances, and, in spite of his many criticisms Mawson was to write down, his admiration for David having done as much as he did was clear – but not at first.

On the fifth of October Mawson had written "the professor dog tired all day as he had not been to sleep the night before". The following day he wrote "the professor doggo again" and again the day after.

On the eighth of October Mawson spelt out his frustration with David : "Prof broke attachment of sledge meter this evening in rage when camping. Prof finds it necessary to change his socks in morning before breakfast, also has to wear 2 [pairs] per day. And comes in late for [sleeping] bag and sits on everybody. God only knows what he does.

"He is so covered in clothes that he can hardly walk and hardly get into bag – that is to say, hardly leaves any room for us as he very nicely made us take side places. He wears at least 1 singlet and 1 shirt, Jaeger wool waistcoat, waistlet sweater, blue coat and burberry, blue pants, double sealed burberry pants, fleece balaclava and fleece lined helmet, burberry helmet."

He did write the next day that "Prof struck better form" but it was a short lived improvement.

Mackay had rigged up the tent floor cloth for possible use as a sail for the sledge and on the fourteenth of October Mawson wrote "Prof very slow so that Mac and I rigged sail and started without him" Later that day when they had finished their man-hauling, Mawson wrote that "An hour or 1½

hours after we had retired the Professor came in and woke us up in his efforts on retiring. He, though absolutely doggo for the last 1 mile march this afternoon, had walked over to the mainland and geologised."

Three days later they reached a low rocky promontory which had been named Cape Bernacchi. Although several of Scott's expedition had visited part of Victoria Land – the western part – and Sir James Clarke Ross had claimed Victoria Land for the British Empire over fifty years previously – David recorded that "on Saturday, October 17. Mawson, Mackay and I landed at Cape Bernacchi, a little over a mile north of our previous camp. Here we hoisted the Union Jack just before 10. a.m. and took possession of Victoria Land for the British Empire" [HoA,2 p 94].

Up till then David himself had been leading the party out in front but two days later he realised that he should hand the duty over to Mawson. David wrote [HoA,2 p96] that on the nineteenth of October he had been struck by snow blindness through not wearing his snow goggles regularly and he then asked Mawson to take over his position at "the end of the long rope, the foremost position in the team". He later wrote that "Mawson proved himself on this occasion and afterwards so remarkably efficient at picking out the best tracks for our sledges and steering a good course that by my request he occupied this position throughout the rest of the journey".

THREE MEN IN A BAG

The next day Mawson wrote at length about his irritation at David's behaviour : "We were quite warm last night and have almost forgiven the Prof for keeping us so long before going to bed etc. He generally comes into tent after we are both in bed and spends ½ hour on top of bag arranging and changing things. He sits on our legs and faces alternately. Finally when we have got the chill off the bag, he struggles in all cold and bedaubed with snow. Of course he has the warm middle berth and occupies certainly more than ½ the bag as he wears innumerable clothes.

"His pockets are full of food scraps, specimens, books, Bonza set (tools, blades etc. in pouch) etc. so that there is little left in bag for us. He is so warm then he likes to leave the toggles undone whilst we shiver. The weight of these clothes makes him ill on the march but he cannot see it. There is no getting him to hurry up and partial rows are frequently at meal times, not only on account of this but also, and chiefly, that he is not economical with the oil."

In David's account of their journey in 'The Heart of the Antarctic' there is no criticism of his two companions – perhaps because it was written some

Mackay, David, Mawson at the South Magnetic Pole.

weeks after the journey and David knew it would be published in Shackleton's book. He did, though, make reference to the difficulties of a three man sleeping bag [p127] when he wrote "A three man sleeping bag, when you are wedged in more or less tightly against your mates, where all snore and shin one another and each feels on waking that he is more shinned against than shinning, is not conducive to real rest; and we rued the day that we chose the three man bag in preference to the one-man bags."

On the twenty second of October the three men discussed their progress in the light of the three objectives which Shackleton had laid down for them. It was clear that at their rate of progress it would be difficult, if not impossible, to achieve all three – to do a geological survey of Victoria Land, to spend at least two weeks in the Dry Valley and to determine (and reach) the position of the South Magnetic Pole. And they also all knew that Shackleton had written that the proposed geological work in the Dry Valley was "of supreme importance".

Mawson strongly advocated that they should abandon the idea of reaching the South Magnetic Pole and concentrate on the two other objectives. David was against this and stated that they should concentrate on reaching the Magnetic Pole, even if it meant abandoning the other objectives. Mackay agreed with David.

By the first of November it also became clear that at their present mileage and their consumption rate of food they could not reach the Magnetic Pole in time to be able to return for the *Nimrod* to pick them up. They had given up the prospect of returning to Butter Point where Shackleton had said he would place a food depot for them and then go back to the hut at Cape Royds. They would have to rely on the ship seeing them somewhere along the two hundred miles of coast line and picking them up there.

David suggested that they should immediately go to half rations to the point on the coast at the Drygalski Glacier where they then turned inland, a distance of some one hundred miles. To supplement their rations they should catch and eat seals and penguins. When they were at Cape Royds during the winter Mackay had made a blubber lamp which could remain alight for several hours, using the oil as fuel. They had not taken this with them but instead now made one from an empty biscuit tin. It proved a success. At the Glacier they would leave one of their sledges and go to the Magnetic Pole with just the one sledge and partly rely on seal and penguin meat for food. By these means they hoped they could reach the Pole and return in time.

David and Mackay went to the highest point on the small island where they had pitched their camp and agreed that would be sufficient for a depot.

Hopefully the *Nimrod* would see the flag on it from the sea when they were looking for the party.

In the depot David left a letter addressed to the *Nimrod*'s captain as well as letters to Shackleton and his own family back in Australia. The letter to the ship's captain included a sketch plan taken from an Admiralty chart to show the position of their intended depot (at Drygalski Glacier) where they would plant a black flag and then turn inland to go to the Pole. David wrote that they expected to reach the Drgalski Glacier by the fifteenth of December and then they would go inland. The journey to the Pole from there and back to the coast at the Glacier was, he estimated, some five hundred and twenty miles and would, he thought, take some six weeks so they would not return to the depot until "about January 25. We propose to wait there until the *Nimrod* calls for us at the beginning of February." [HoA 2 p110]

That, of course, presupposed that the *Nimrod* would get back to Cape Royds in January from New Zealand and that Shackleton's own polar party would get back to Cape Royds by the beginning of February at the latest. If all went well then the *Nimrod* could sail along the coast of Victoria Land and find them. It would all be a very close run thing.

While they did manage to take magnetic and geological observations – and Mawson at one point completed a trigonometrical survey of some two hundred miles of coast – their progress was continually affected by biting winds, low temperatures, falling snow, falling down crevasses and, sometimes, blizzards that prevented any movement at all. Altogether, five of the days they were travelling were lost due to blizzards forcing them to remain in their tent all day.

THE WATCH

By the end of October David's behaviour had exasperated Mawson. He wrote [29 October] "The Prof has worried us again by coming to bed in the early hours of the morning uncomfortably cold and persistent. Yesterday morning put the cap on the chronometer which the Commander [Shackleton] had specifically given me to look after, regulate and be responsible for, has exercised the Prof ever since I got it. He was originally sulky over it; since then adopting the celestials' tactics, he has tried diplomacy to wrest it from me. On all occasions he has asked me for the time, especially 3 or 4 times in the early hours of the morning, by saying till I am sick of it, 'Would you mind kindly letting me know the time from your watch presently, there is no hurry, if it would not be troubling you too much, please'. He made it so obvious every time that he wanted the watch that, much annoyed at his roundabout tactics, I gave it to him to look after yesterday on condition he took full responsibility and handed it over to me on completion of the journey."

Mawson went on to make further criticism of David : "He is full of great words and deadly slow action – the more we bustle to get a move on the more he dawdles, especially tying strings to one another and all over the sledges, which have all come off again in unpacking.

"He dodges packing sledges every morning, then, when we are waiting to press on, having packed up, he comes along with a lot of wants and things to be put into the already packed bags. Finally, when all ready to go, he must have a rear" [defecate].

"He will take all day putting roundabout questions to one in order to get a simple Yes or No answer. This worries one almost to distraction."

Mawson then wrote about his disquiet that they were abandoning two of the objectives that Shackleton had laid down for their journey and just concentrating on the South Magnetic Pole. Later in his diary Mawson recalled how originally he had wanted to go as geologist on Shackleton's expedition but David had cabled him and appointed him the expedition's physicist. At that time David had not been intending to winter over but to return to New Zealand on the *Nimrod* after Shackleton's party had been landed. Now that had changed. David was to stay with the shore party and not go straight back to New Zealand. Mawson had then met Shackleton at Christchurch in New Zealand and he had confirmed David's appointment of him on the expedition.

As a geologist Mawson had wanted to prospect for minerals, survey along the coast of Victoria Land and particularly do survey and geological work in the Dry Valley. This was a unique part of Antarctica, with no snow or ice, found by Scott during the *Discovery* expedition. When they were on their way Mawson reckoned that David, all along, had regarded the South Magnetic Pole as the main objective of the party and as Mackay agreed with David, he had to go along with them.

For the next three weeks Mawson made no reference to David's eccentricities in his diary. Then, on the twenty third of November he wrote : "The Prof is certainly a fine example of a man for his age – he does more than any other man of his age could – but he is a great drag on our progress. He certainly and admittedly does not pull as much as a younger man."

Mackay's diary starts at the end of November – oddly Mackay had listed his first entry as on "Nov. 31st but it must have been the thirtieth of that month. They were approaching the Drygalski Glacier with inlets on the coast, a range of small mountains nearby, a jumble of sea and coastal ice, ice ridges and crevasses and no clear way to go forward.

"Nov. 31st

First day on Drygalski barrier. Curious formation, very similar to Culbin Sand Hills. Rounded ridges of ice about 100 feet high, running about N & S crossed by ridges of snow at right angles. Plateau wind has been blowing all day, but it has now dropped, and thermo is at +28°. Very hot inside tent. Our working day now is of course from midnight to noon.

"Dec. 1st

Hauled sledges on about one mile towards middle of barrier. Found country growing rougher, so halted, lunched and then Mawson took dip. observations, while prof. and I went ahead reconnoitring. Country continued to grow worse. We decided it was impassable and resolved to return to southern shore of barrier, travel outwards till barrier appeared smoother, and then cross. No plateau wind today, very warm.

"Dec. 2nd

Retraced our steps to edge of barrier, and then turned Eastwards, made a good march on excellent surface. I spent a sleepless night thinking the others were inclined to give up, but this morning they both declared themselves keen. Of course we are all stale, but otherwise fit.

"Dec. 3rd

Marching along south side of barrier, snow fairly good, but deep and rather soft, making walking very tiring. Did 3 ¼ miles. Prof. and Mawson reconnoitred barrier after dinner, and think it practicable. It certainly looks smoother than at the last place we tried.

"Dec. 4th

Hauled sledges half a mile on to barrier, made an early lunch, and then started, all three to reconnoitre to the Northward. Penetrated about 5 miles in to barrier, but could not be certain that we saw N. side. Made out a route just possible for sledges, but the crossing will prob. take us about a week.

"Got back to camp at noon for lunch, all very tired. The Prof. now asked me if I would go back to the last place we had seen seals and get some meat. I offered to do so, and started."

David put this somewhat differently in his narrative [HoA p140] when he wrote that because they now had only one day's supply of seal meat left "I proposed to that evening out to the berg in search of seals, but Mackay kindly volunteered instead."

SEAL MEAT

Mackay's diary continued : "It must have been about six miles there and six back in soft snow, and took me twelve hours of continuous walking. So that was over 24 hours going. I got lots of seal meat, and one Adelie penguin. Another one walked into camp and was killed by Mawson. The bag which I carried back must have weighed 40 to 50 lbs." In fact their journey would have been impossible without killing and eating seals and penguins – and it was Mackay who was the main executor of that task.

Mackay was exhausted when he returned to their camp. Mawson wrote that Mackay was "doggo on return". David wrote that Mackay "by securing the so much needed additional supply, he had rendered us an extremely important service." The fifth of December they spent in their tent. Mackay's entry for that day was "Dec. 5th Spent in eating and sleeping."

"Dec. 6th

Advanced 3 miles, in a devious course, though ground is not very bad. Made about 1 + miles to the good. Rations a bit short.

"Dec. 7th

Very like yesterday. Sky was much overcast, making it very difficult to see our way, but we made good travelling. In the afternoon a strong southerly wind began to blow, and clouds cleared away. We are on short rations both of biscuit and seal meat. Our biscuit now runs to one for breakfast, 1 + lunch and + dinner, but all these measurements are short.

"Dec. 8th

Very heavy pulling through deep, soft snow. Fine view of Mt Nansen range, and pass or glacier to south of it, up which we are to travel to reach the plateau. Glacier appears rather rough, I am afraid.

"Dec. 9th

During a reconnaissance today, I saw the whole of Terra Nova Bay clear of ice, and a long stretch of the barrier edge, with what may be called the Nansen barrier beyond. The barrier edge appears about 5 miles off. All flat going ahead, but if the surface is anything like what we have passed over today it will be the worst of pie-crust snow. The professor saw a Wilson petrel. I saw a flock of half a dozen snow petrels and three skuas. Mount Melbourne is smoking actively.

David wrote [HoA 2 p143] that on that day Mackay, going ahead with field glasses to reconnoitre, had spotted open water on the northern edge of the Drygalski ice barrier at the foot of the glacier, some three to four miles distant, and had shouted "Θχλχττχ Θχλχττχ" which thrilled us now as of old that thrilled the Ten Thousand. This was a reference to the account by Xenophon of the shouts of the ancient Greek army, when travelling back across a hostile Persia, had at long last seen the sea. Mawson briefly mentioned this in his diary two days later. Mackay had not mentioned this at all.

"Dec. 10th

We seem to have got out of the rough stuff, and are now heading for Mt Melbourne. Although we have ascended several hillocks since the one from which I had my view yesterday, we have not had nearly so good a one again. I estimate the barrier edge to be about 10 miles off. Our depot to be at about 20. Rations are very short, and I am hungry. I feel as if we had very little chance of the pole.

"Two or three days journey will show us."

From now until they were eventually picked up by the *Nimrod* they were constantly concerned about whether they would make it to the Pole and get back to the coast in time.

"Dec. 11th

Really a joyful day. Marched 3 miles 1000 yds. Camped within about a mile of the sea and 5 or 6 miles of the foot of Nansen glacier. Everything seemed to jump closer to us this morning. The Nansen glacier looks good going, icy, and not very rough. The low sloping shores marked on map appear to be a mystery or myth. Lots of seals in sight."

A NARROW ESCAPE

David wrote that on that day he had had a very narrow escape. While Mackay was looking for the best way forward and Mawson sorting out his camera plates, he decided to go out with his sketch book and draw an outline panoramic view of the coast ranges then in sight. He thought that because they had had little trouble with crevasses in the immediately preceding days he would not take his ice axe : "but I had scarcely gone more than six yards from the tent, when the lid of a crevasse suddenly collapsed under me at a point where there was absolutely no outward or visible sign of its existence, and let me down suddenly nearly up to my shoulders. I only saved myself from going right down by throwing my arms out and staying myself on the snow lid on either side. The lid was so rotten that I dare not make any move to extricate myself, or I might have been precipitated into the abyss." [HoA 2 p145]. Fortunately Mawson came out of the tent and with his ice axe, dug a hole in the firm ice and pushed the haft to David. He grasped this and then managed to climb out. Neither Mawson nor Mackay mentioned this in their diaries.

"Dec. 12th

Shifted camp on about a mile, then after a long confab between Mawson and Prof. decided we might as well form our depot here, on top of an ice knoll. I was sent off to kill seals and penguins to be cooked for the plateau trip. Worked along where sea ice is marked on sketch chart" [Mackay had drawn sketch maps of the area where they were travelling] " and killed six seals, three Emperors and one Adelie penguin. It was rather disgusting work. The sun was very warm, I never felt cold, and my clothes, all but my boots, dried on me.

"Dec. 13th

Spent in making small repairs, and eating. But the Prof as usual will not let us sleep enough."

It was now that they built a depot with one of the sledges in it, put food and some of their equipment also inside and letters to their families and David's letters to the *Nimrod* captain and to Shackleton. They fastened a black flag to the depot and loaded the other sledge with provisions for the journey to the Pole. They were now going inland with the other sledge and leaving the coast. Altogether they were taking provisions for seven weeks, a total – with their tent, sleeping bag and equipment – of some six hundred and seventy pounds on the one sledge.

"Dec. 14th

Eating my best and writing letters. These are last adieus, so they ought to be tragic, but I cannot make mine so, as I feel we have such a good chance of reaching the pole. Fixed up our depot finally.

"Dec. 15th

Woke to find a blizzard blowing from the plateau. There are no signs here of the south-easterly blizzards that we used to have at the hut. I was glad to keep to the bag, as our stay here has not been much of a rest. Two front poles, by getting covered with rough ice, very nearly wore their way through the tent. Today we started our full rations, with seven biscuits a day."

SAVAGE CRITICISM

During their journey the three men had taken it in turns to cook the food each week. Although Mackay had been the one to get more seals and penguins for eating than the others, and though it was his design which helped to make the blubber stove, Mawson was not impressed by his efforts. In a savage criticism in his diary for the sixteenth of December he wrote : "Mac was very lazy at depot and very unskilful [sic] in cooking and generally in everything but hard plain manual work. He would make a good soldier but no general." It was, though, the only criticism he would write of Mackay.

Mackay's next few diary entries were quite optimistic but were not to remain so for long.

"Dec. 16th

Still blowing in the morning, but moderated about 5 a.m. We struck camp and started with our single sledge at 7 a.m. and by 10 a.m. had done 3 miles 1500 miles, at which I, at any rate, was pleased.

"Dec 17th

Run of 9 miles 100 yds over level barrier surface, sometimes very soft. Crossed several cracks, in one of which sea-water was showing. Barrier appears to be not more than 20 feet above sea level. At present we are camped in front of a broken barranca some 20 or 30 feet deep and 100 yds wide.

"Dec. 18th

Run, 9 miles 350 yds. Good surface but several large undulations across our tracks, almost a mile from crest to crest, and I suppose about 100 ft high, so it was a pretty good performance. These waves in the barrier are caused, I suppose, by pressure from the glacier, and are roughly concentric round its mouth. We seem to be still about 5 miles from glacier foot, though I thought that our depot was at that distance. We passed a considerable moraine [ridges on deposits of rock debris transported by a glacier] outcrop in the middle of the barrier.

"Dec. 19th

Day began with snow, fog and wind. No land visible. Started by compass due magnetic south. After passing over two undulations, came to a sheet of flat ice with tide-cracks, showing open water, which tasted salt, ice at edge of cracks apparently 1 ft thick. Whole thing most puzzling. Passed off this ice up very steep incline on to another large ice wave. Much crevassed. Mawson fell right into one out of sight and it was a job to get him up. Lunched soon after. Fog lifted a little, but settled again. Made a short reconnaissance and then dined and turned in. It is blowing and snowing, and looks bad for the pole."

David described the incident in which Mawson had fallen down the crevasse in some detail [HoA 2 pp155/6] : "Fortunately the toggle at the end of his [Mawson's] sledge rope held, and he was left swinging in the empty space between the walls of the crevasse, being suspended by his harness attached to the sledge rope. Mackay and I hung on to the rope in case it should part at the toggle, where it was somewhat worn." Mackay then threw down the end of the alpine rope they had and, by putting his right foot in the bowline which Mackay had tied, he was lifted up and finally got out on to the ice. Mawson merely wrote in his diary that "I fell into one [crevasse] but hauled out with aid of alpine rope".

Mackay continued :

"Dec. 20th 5 p.m.

Up at 8 a.m. Foggy. Determined to abandon attempt on Nansen glacier. Though I voted against this. Started south easterly skirting round Mt Larsen. Ice is undulating and crevassed and there is six inches of soft snow on surface which conceals crevasses, and sometimes jams the sledge altogether. At 4 p.m. a blizzard at about temp of +32°F sprang up from the E.S.E. Stopped and camped. For the last few days we have been much troubled by the dampness of everything, due to temperatures above freezing point."

David wrote that at this point the prospects of reaching the Magnetic Pole and getting back to the depot at the Drygalski Glacier were not good. They reckoned they had to be back there by the beginning of February if they were to be picked up by the *Nimrod* and they had not yet climbed up from the mass of ice, ridges, glaciers, mountains and crevasses on to the plateau from where they could go straight to the Pole. Furthermore the soft snow made pulling the sledge very difficult.

The next day Mawson wrote further criticism of David and made several such entries in the days ahead. Clearly David was becoming a problem

Mackay wrote :

"Dec. 21ˢᵗ

Up at 2 a.m. Fog cleared away and we resolved to reconnoitre towards foot Larsen glacier. Roped up and walked in that direction, but found the ground broken by pools of water, ice-ridges, crevasses and snow drifts more than a foot deep. Returned and lunched, and then reconnoitred up a snow slope apparently curving over a spur of Larsen and leading on to Larsen glacier. Reached a height of 1500 feet. Slope far too steep, and snow 18 inches deep and very soft. But must try it. On the way back, the noise of running waters in every direction was quite loud and we heard several considerable streams in the ice under our feet. Got a good view of Nansen glacier, which I am afraid is too rough to be practicable."

SNOW BLINDNESS

Mackay then had a bad attack of snow blindness and wrote nothing for the next two days. Mawson wrote on the twenty first of December that "The Prof was doggo this afternoon. I had him well under observation and showed him to be in nothing like the condition we are in".

Mackay started his diary again on the twenty fourth of December.

"Dec. 24ᵗʰ

Had a bad attack of snow blindness, result of doing our reconnoitring without snow spectacles. It really was most painful. We are now camped 800 feet up, on our snow slope, having tackled it with half loads. We found a good route in to the foot of slope, but directly we got there a blizzard from the plateau sprang up. We had just time to get tent up when it was on us hard, and blew for 24 hours harder than anything I have ever experienced in a tent. It ripped the tent in two places, and split the peak of it. So this morning Mawson and I patched it, which was very cold work. We found the whole tent very rotten, and I don't think it will stand another blow such as we had. The blizzard has done good though by clearing the soft snow off the slope, and leaving large patches of bare ice. But for this we could never have got up at all. We have got things up the glacier so far in half loads."

On Christmas Day Mackay wrote "no Christmas luxuries at all". David, though, wrote in his narrative that he and Mawson gave Mackay some sennegrass to smoke in his pipe. It was the only Christmas present they could give.

The same day Mawson wrote "Prof very doggo" and "He has of late appeared to have lost all interest in the journey". David was getting worse – he could not pull as well as the other two, he seemed listless and his manner became more eccentric. In his narrative, which admittedly was written later after the expedition was over, David makes no reference to his own behaviour or discomfort.

Mackay's entry for Christmas Day continued :

"Dec. 25th

It was blowing very hard in the morning and so after breakfast we got into the bag again till the wind went down at about noon. Then started up our glacier with fully loaded sledge. Did 3 miles and reached a height of 2000 feet. Our blizzard glacier opens on to the one coming down between Larsen and Belinghausen at this height. Our camp is on the middle of this glacier. Going good. Temp. about +25, wind about 15 miles an hour.

"Dec. 26th

Did 8 miles and rose to 3280 feet. I am well pleased, as surface was spoiled in afternoon by a light fall of snow. The clouds have been rolling about, spoiling the view, but sometimes producing beautiful effects. They are coming in from the sea, with a very light N.E. wind. Temp. +26. We must be situated somewhere at the back of Mt Larsen and can see, down on our left, what I think must be the Drygalski glacier.

"Dec. 27th

Run 10 miles. Altitude 4050 feet. Temp. from +7° at 8 a.m. to +23° at 2 p.m. Slight N.W. wind. Sunny generally, clouding over more in afternoon. I remark that these clouds, above and to the west of us, move very very slowly, taking hours to alter shape. One large one straight in front of us seemed as if it were leading us to the promised land.

"The day's run is good, as we did not get started till 11 a.m. We spent some hours forming a depot of ski boots, ice axes, alpine rope and a few odds and ends, lightening the sledge about 24 lbs. We really left our glacier this morning, and have been fairly on the plateau all day. We are dipping Mt Larsen behind us, and are opening out some old friends, Mt Bowen, Howard etc. to the Southward."

They named the depot 'Larsen depot'. It was a balance between taking sufficient food for the journey to the Pole and back and reducing the load on the sledge so they could travel a few more miles each day.

"Dec. 28th

Run 10 miles. Altitude 4650 ft. And we had to dip into a hollow around 100 ft deep, caused by head of Nansen glacier. Besides this we stopped for 1 hour after for Mawson to take dip observations. So I am pleased on the whole. We are not losing strength. M. is afraid that his observations make the pole farther than he had placed it, that is 170 miles from here. He cannot be certain, however without more observations. The day began very warm, +23 and completely overcast, but now, 8.20p.m. it feels much colder. We are, I suppose, about 20 miles south of Nansen, and have a splendid view of it, and the tail end of what is probably the Mt Baxter range. We can still see Larsen to its base – that is, where it rises from the plateau, but we lost it when we were in the bottom of the hollow. Mawson now says that the nearest the pole can be is 170 miles and the farthest 230. If the former, we can do it, if the latter, we may possibly."

The same day Mawson had written that Mackay had admonished David for not pulling his weight with the sledge but Mackay himself made no mention of this. The irony of the diary entries is that while Mackay was a fairly short tempered man and Mawson was much more even tempered, it was Mawson who wrote about almost all the frustrations and worries about David.

The next day Mackay was pleased with their sledging.

"Dec 29th

Run 11 miles!!! Grand! Alt. 5280 ft. Day began cold, at +7 at 8 a.m. with a breeze. D – d uncomfortable. Breeze freshened and lasted all morning. Mawson took a meridian altitude, which tended to confirm opinion we have only 160 miles to do. Mt Nansen in full view from behind, and we go on opening up peaks to the North of it. Mts Larsen, Belinghausen, Neumayer and Bowen seen at intervals, we lose them, when we dip into a hollow, which may be otherwise quite imperceptible. Feeling the exhaustion and hunger awfully, but the less said about it the better.

"Dec. 30th

Run 11 miles, 5900 feet. Temp. 8 a.m. Rose to +7, never higher, and breeze until mid-afternoon. In the middle of this ripped a hole six inches long in tent. Had to patch it in the breeze. It was intense torture. We have opened two more peaks to N.E. Don't know their names. We also saw a short, hog-backed, crevassed ice ridge, about 5 miles to N. of our course

"Dec. 31st

Run 10 miles, 6500 ft. Temp. 10 a.m. – 1. Max. +9. Almost calm all day. But we all agreed that none of us ever felt a day's pulling harder. I was nearly dead. M took a dip observation, which makes the pole farther off than ever. I hope we are not growing weaker. We had some steep hills to go up, and there are steeper ahead. In fact there is a crevassed ice ridge on our left front, running across our course, but pretty smooth straight ahead. The tent needed more repairing. Luckily the weather was calm. A skua came down beside us as we were pitching camp!"

In his entry for that day Mawson exploded over David's behaviour. He wrote: "The Prof is dreadfully slow now, he does nothing. Mac mends tent, I mess, he is always still sitting down when we are packed and pushing our way out of tent. This morning I told him he was keeping us waiting as he was not attempting to get ready and we were all packed after working. He looked very angry at my saying this and started packing up loose impedimenta, then went out of tent, had rear, and did up his wineskins. In the meantime we had tent down and sledge strapped. He never, or seldom, helps pack a sledge – even at lunch time he is content with looking on. Something has gone very wrong with him of late as he almost [always] morose, never refers to our work, shirks all questions regarding it, never offers a suggestion. Well anyway, he is getting the value of our blood as we (Mac and I) do our level best at pulling and generally pushing on the expedition"

Mackay again did not mention David's behaviour and it is unclear whether at that stage he and Mawson actually discussed it.

NEW YEAR 1909

Mackay's diary continued:

"Jan. 1st

We all wished each other happy New Year, and we ought to be happier than we were at Christmas, for we have a much better prospect of reaching the pole.

"Run 10 miles. Altitude 6980. Temp +6 to +17, no wind. The snow slopes did not prove so steep as we expected, but the surface generally has been very soft, making the pulling very hard. A sort of rumour started that the meter was under-registering, to the extent of only showing 3 miles for 4 covered, but I don't believe in it. Mawson is giving us a special thick hoosh, by way of a New Year's day dinner. I have just finished it, and could easily eat three more of the same. Of course our hunger is simply agonising. Well, we can't have more than a month of it now. The Admiralty mountains are hid in mist. 62 miles from Larsen depot.

"Jan. 2nd

10 miles, 7250 ft. Temp. pretty steady about +8. Little wind. A dreadfully heavy day, with bad pie crust surface and several undulations. It is all we can do, under such conditions to keep up our 10 miles. We are all absolutely exhausted, and I am afraid growing weaker. The undulations make me anxious, as of course they will tell against us coming back. I see, on looking back that I have not yet explained that our plan is to go on at 10 miles a day till the 15th, then turn, hoping to come back at 15 miles a day. At present I think we ought just to be able to do this. Prince Albert Mts, especially one peak which we think must be Mt Queensland, are showing up well, the plateau sloping towards it. But we have lost Mt Larsen and can only see the top cap of Mt Nansen. 72 miles from Larsen depot.

"Jan. 3rd

10 miles, 7810 ft. Temp. +8 to +1 [We] did not get into camp till 9 p.m. after the heaviest day's pulling we have had; larger undulations and softer snow. Mawson took another meridian altitude, which makes it certain that we are only doing 10 miles a day, the meter being correct, and he having made a mistake in working out his last sight. We can now see 100 miles or so of snow covered plateau stretching away N.W. of Mt Queensland, but without any high or conspicuous peaks. In fact, it surprises me a good deal that the mountains that appear so rugged and irregular from the sea, should form such a smooth, continuous wall to the plateau as they do on this side. We can only be sure of one pass or gap through this wall, which is seen distinctly to the south of Mt New Zealand. Our hunger is too dreadful to speak of, but it is not for more than a month now. Twelve days, and we will be scudding back all we know. 82 miles from Larsen.

"Jan. 4th

10 miles, 7850 ft. Temp. -5 to +6. Day began with a stiff breeze which blew strong all morning, but died down soon after lunch. Surface improved slightly

after lunch, and it is almost dead level. We all complained much of exhaustion at lunch, but as it was Mawson's last day of cooking he managed to give us a little extra biscuit, so we pulled hard in the afternoon. Tent tore again, and I patched it, but no wind, so not very cold.

"Sky is now almost completely overcast, the first time we have had an overcast sky. No landmarks visible. 92 miles from Larsen.

"Jan. 5th

10 miles, 7950 ft +17!! Overcast and calm. Quite level and fairly good surface. Not feeling so tried. I am cooking now, so no time to write much. This is half way from Larsen depot. 102 miles."

For the next eight days they went on averaging ten to eleven miles a day and it looked after all as if they now would make it to the Magnetic Pole and be able to get back to the depot at Drygalski Glacier and the coast where the *Nimrod* might see them. It was arduous work, though, and cold with strong winds and not enough food.

Mackay's diary continued:

"Jan. 6th

10 miles, 8000 ft -4 to +7 Slight blow from S.W. Calm and warm in evening. Not so tired, but sick of the whole show. Surface fair. 112 miles. Suffering much from split lips.

"Jan. 7th

10 miles, 8700 ft -13 to -3 and a stiff breeze most of the day with very bad surface. I never felt so exhausted and hungry. Mawson took sights, which tally with cyclometer record. 80 miles to go, and 122 miles from Larsen. Mt Queensland was in sight at lunch, but we have lost it this evening. The cold is becoming trying. 122 miles.

"Jan. 8th

10 miles 8900 [feet] -15 to -10. A most dreadful day, a strong blizzard almost abeam, and very bad surface. I never felt in lower spirits. We tore the tent badly in putting it up for lunch and had to patch it while it was up, after dinner. All clothes which were out drying got covered with snow drift, and my hands went completely several times. Pray God we may never have such another. We marched right through it. 132 miles.

"Jan. 9th

10miles. 9000 ft. -3 to -7 calm, and surface better, more cheerful all round. 142 miles."

It was on the ninth of January 1909 that Shackleton, Wild, Adams and Marshall reached latitude 88° 23' south, longitude 162° east : the furthest south any human beings had ever been and, indeed the nearest anyone had ever got at that time to either the North or the South Poles. It has always been doubtful whether Shackleton really thought he could actually get to the South Geographical Pole but certainly his dream for his expedition was for his parties to get to both the geographical and magnetic Poles. And he almost succeeded.

Mackay's diary continues :

"Jan. 10th

11 miles 9000 ft. Temp. -7 slight southerly wind, taking us abaft the port beam. Mawson took sights, which confirm our position as shown by compass and cyclometer. The 11 mile run was not all done today, but was the sum of several daily increments. All cheerful. 153 miles.

"Jan. 11th

11 miles, 9000 ft. Temp. -12 to -5. The wind blowing fresh on port quarter, and a fair surface. Also, I believe, a slight slope in our favour. I don't like all this, as it makes me anxious about getting back. 164 miles.

DINNER PARTIES

"Jan. 12th

11 miles, 9000 ft. Temp. -15 to -3 but such a bright sun and so calm as to feel quite warm. Day began cold, however, and with a fairly dense fog which soon lifted. There was no wind, but the sastrugi although confused, are on the whole, in the direction of the last two days wind. This is very interesting. In weather like this, sledging is 'sledging de luxe' but for the awful hunger, which weakens us all, and has a bad effect on our tempers. Today, during halts, we passed the time by planning menus for two dinners, a Scotch and a Yorkshire, to be given in Sydney."

Here Mackay spelt out his two menus. Mawson listed them as 'Professor's Menu for Yorkshire Empire Dinner' (Mawson was born in Yorkshire before his parents emigrated to Australia) and 'Professor's Dinner to Mac (Scotch Dinner)'.

Mackay concluded in his dairy entry for that day : "It is wonderful what a lot we think and talk about our bellies. I could almost eat my finneskoe. We should be at the pole now in three days. The compass still acts, though very sluggishly. 175 miles."

That evening of the twelfth of January, when they thought they were nearly at the Magnetic Pole, then saw Mawson make a dramatic announcement and one which threatened their entire journey. The Magnetic Pole was further away than he had previously estimated. That meant they had forty more miles to go than they had anticipated. Given the small amount of food they had and their severely weakened state – let alone what would happen to the weather -could they possibly do it?

Mackay's diary recorded his feelings at Mawson's announcement.

"Jan. 13th

13 miles. 9000ft. Temp. -6 to +2. Overcast almost all day. Calm, or with puffs from South.

"Last night Mawson made the astounding announcement that the pole is prob. 40 miles farther off than we had ever thought. He is led to think this from re-reading the *Discovery* "reports" and from his own observations. We were left to think over this during the night, and in the morning, after a very heated discussion, we determined to take it on. I, of course, agreed to go if the others were decided, but I said plainly, as I think now, that we have not more than a 50 per cent chance of getting back to the coast in time for the *Nimrod* to take us home. My reasons for thinking so are : That the Professor is very nearly crocked now, and we are both weak. That we have no reason to suppose that the first half of the journey back will be a bit easier than what we have done, since any wind we have had during this part of the journey has been abeam or astern of us, and the sastrugi point to this being the prevalent wind; and last that we are making no allowance for weather so bad that we cannot travel. So that is how I feel.

"At present, Mawson promises to turn the sledge homewards on the morning of the 18th and hopes to make an average run of 13 miles a day, thus getting us to the sea on the 7th. I feel little the worse of our 13 miles today. But the Professor looks quite crocked up.

"Mawson made the pole 50" distant by dipping circle at noon. That is about 45 statute miles. But it was a rough reading. 188 miles."

In his narrative David merely says that after discussion they decided to go on for another four days. Mawson, though, recounted the details of their discussion. He stated that David wanted to go but while Mackay said he would be willing to go on – four extra days and four more back – he did not think they would make it in time. At last Mackay apparently agreed to the extra journey providing they would travel thirteen miles each day. "Mac" wrote Mawson "wants us to swear we will do 13 m per day return ; it is done."

In fact, in three more days they got to where Mawson calculated the South Magnetic Pole lay and where in accordance with Shackleton's instructions, they hoisted the Union Jack on the flag pole David and Mackay erected, and where David said "I hereby take possession of this area now containing the Magnetic Pole for the British Empire" [HoA p181].

Mackay wrote in his diary :

"Jan. 14th

12 miles, aneroid unreliable. Temp. -6 to zero. Perfect weather, blue sky with Noah's Ark clouds. Wind very light, but hauling more aft on port side, as also are sastrugi. I have an impression that we have been going down hill, and cannot help feeling anxious. Surface has been much roughed, evidently ploughed up by violent winds. I feel the pinch in my belly very bad. 200 miles.

HALF RATIONS

"Jan. 15th

14 miles. Altitude? Temp. -20 to -6. Southerly wind of 12 miles per hour, sky perfectly clear. Mawson took dip at lunch, got 89* 45' which he says mean we are about 30 miles from pole [sic], which I cannot understand. Anyhow tomorrow is to be our turning day. We are to take the sledge 8 miles and walk 5, but I shall describe later. We have this evening 22 days rations left, on our twice diminished scale, namely : six biscuits a day, and about two thirds of a mug full of solid pem [pemmican] among the three of us twice a day, with tea, cocoa, milk and sugar. It sounds not so bad, but is really little more than half rations.

"If weather holds out, and we can keep up 13 miles a day, we will get back all right. If not, then God Help us. 214 miles

"Jan. 16th

Up at 4.30. Took sledge on with full load 2 miles, then off-loaded everything but tent, sleeping bag, two day's food and a few clothes. Sledged on other 6 miles, camped, lunched at 10.30. We are going to walk from here another 5 miles with a compass, to where Mawson thinks the pole must be. Last night he took a dip-reading, giving him 0° 12' off the perpendicular.

Just returned from magnetic pole, five miles to N.W. of our camp here. We hoisted a Union Jack there, and the Professor in a loud voice, annexed the place. Mawson photographed the three of us by means of a string attached to camera. Then it was only at my suggestion that the Prof. called for three cheers for the King. At the pole, the compass still pointed, very sluggishly, towards the N.W.

"The weather perfect, bright sun, but with slight wind from South, which plays Old Harry with my lips. The lower one is now almost completely bare of skin, and so split that my mouth has a three cornered appearance.

"But we are all in the best of spirits.

"9 p.m. Camped at off-load, that is a days run of 24 miles. None of us are crocked. The afternoon has been absolutely glorious, a flat calm, with hot sun. We are talking of doing 15 miles a day on the road home."

From the location of the Magnetic Pole they had two hundred and sixty miles to go before they reached the inlet at Drygalski Glacier where they hoped the *Nimrod* would pick them up. Shackleton's instructions to David were that if they had not got back to Cape Royds by the first of February then the *Nimrod* – assuming that the ship herself had got back to Mc Murdo Sound from New Zealand – would travel as close to Victoria Land itself to look for them and be "on the look out for a signal from you flashed by heliograph" : but he made it clear the ship would not proceed beyond Cape Washington, just beyond the Drygalski Glacier.

It would be touch and go if they got back at all – and, especially, if they got back in time for the *Nimrod* to see them.

They were all exhausted and, in spite of some optimistic jottings in his diary, Mackay and the others were getting on each other's nerves. Mackay wrote :

"Jan. 17 th

(Glypsometer reading in morning 96.7) Run 16 miles, Alt.   Temp. -2 to +2.

"Camped on site of our camp of 14th, that is 200 miles from Larsen. Weather very warm, a complete calm, or very faint wind from W.N.W. God is very good to us.

"I am greatly pleased with the day's work, the more so, as the surface is not particularly good, and we all think we have been running up hill, though the aneroid is suffering from a contusion, and Mawson won't run the hypsometer.

"Our outward bound tracks are clearly visible. We talk of reaching the coast on the 3rd.

"Jan. 18th

Run 16 miles. Temp. -17 to -3 Alt. ? Very light southerly wind, some high clouds lifting from S. I believe we could have done another mile or two, but Mawson complained of pain in his leg, and the Professor was utterly crocked. I stupidly lost my warm pyjama jacket off the sledge, so I hope we will be down in warmer weather soon. Our old tracks (five days old now) are still plain, and of course we follow them. This saves M. a great deal of trouble in navigating.

TWO FOREIGNERS

"The Prof. (cook for this week, ending today) has been able to give us three or four tremendous fine hooshes, which have an undoubted effect on our pulling. 184 miles from Larsen.

"Jan. 19th

Run 16 miles. Temp. -12 to -5. Alt. Southerly wind of 12 to 15 miles an hour blowing most of the day, with drift at times. Very soft surface part of the way, and some up hill. I don't think I have ever felt so utterly done. At the end, I couldn't stand for half a minute or so. But we will seldom have more trying days. Mawson put some sugar in the hoosh this morning by way of experiment, and I am sorry to say, I lost my temper with him so far as to say he was selfish. He objected to this, so I apologised. We all felt the cold very acutely this morning, and through the day, far more so than the -17 or -19 which we have had on fine, still mornings. 168 miles from Larsen."

Mawson did not refer to the incident with the sugar in his diary. David did mention it in his narrative and said that Mackay had exclaimed that "this awful state of affairs was the result of going out sledging with 'two foreigners' " [HoA2 p 185].

"Jan. 20th

Run 16 miles. Alt. Temp. -20 to -6. Wind pretty steady on starboard beam and tending to haul aft, very slight drift. Surface rather soft, and we have quite left behind the large sastrugi of 17th and 18th. I am not quite so tired as last night, and I am pleased at doing 16 miles, in spite of having had such a tiring day and only seven hours sleep after it. We are all feeling the cold badly, so Mawson proposed at lunch today, that as we are keeping up the pace so well, we should increase rations. This was agreed to, and the result is a most splendid hoosh tonight. We are now on ground covered on the outdoor journey on the 10th, and can still see our old tracks faintly. One awkward discovery is that we have no tea ration for this week, so we pick up the old used muslin bags of tea at the old camps as we pass, and use them again.

"Jan. 21st

Run 16 miles. Alt. by hypsometer 196.7 deg. At lunch. Temp. -20 to -3. Wind as yesterday, but rather lighter and we are feeling a bit warmer, partly by reason of the increased rations, no doubt. Also I don't feel so horribly exhausted and inclined to vomit up my food, as I have done for the last two or three days. We are now getting seven biscuits a day.

"Surface was rather bad for the last 3 miles, a good deal of 'pie-crust' snow. We hope to have the wind a little abaft tomorrow. Our tracks of the 8th (the blizzard day) are still visible, which strikes me as most remarkable, the more so as camping grounds are pretty well snowed up, and we have not been able to get any more tea bags, they being buried by drift snow. 136 miles.

"Jan. 22nd

Run 15 miles. Alt. by hypsometer 196.75 at lunch. Temp. -20 to -3. A dreadfully hard day. We had to stop at fifteen miles as it was growing so late. I am in pretty fair agony for want of sleep, as we only get 8 hours in the bag, and half of that time is spent in shivering. The strain of the whole thing, the exhaustion and actual muscular pain, the cold, the want of food and sleep, the monotony, and the anxiety as to what is to happen at the end, make me think that this must be the most awful existence possible. My thumb nails are both coming off from frost bite.

"We lost the old tracks after lunch today, and almost at the same time I picked up Mt Queensland, which is most opportune, as it gives us a point to steer by. The surface has been utterly shocking. 121 miles out."

Mackay's description of their plight is both vivid and painful. He was not the only one to feel such anguish. Mawson's legs were very stiff. Their lips were badly cracked. Mackay's and Mawson's eyes needed treating for snow blindness. David must have felt it worse than the others. He was pulling less and getting weaker.

It was no surprise that with the increasing anxiety and stress that tempers would be fraught. But the next few entries in both Mawson's and Mackay's diaries did not mention any arguments or criticisms. They were determined to press ahead as fast as they could.

"Jan. 23rd

Run 16 miles, Alt. 196.78 (hypsometer at lunch) Temp. -19.5 to -6, but the sun felt brighter and warmer. The wind was on starboard quarter, but fell almost to calm after lunch. Surface fair. No great drop in altitude. The aneroid readings at this stage of the outward journey appear to exaggerate the height. I blistered my heel badly owing to freezing of finneskoe. 105 miles out.

"Jan. 24th

Run 16 miles. Alt. by hyps. At lunch 197.68 deg. Temp. -15 to -7. Day began with strong S.W. wind, very cold. Surface being fair, we did almost 3 miles first stage, but things got gradually worse. In afternoon surface was as bad as we have ever had it. 89 miles out.

"Jan. 25th

Run 16 miles. Alt. by hypsometer at lunch 197.7. Temp. -15 to -6. A most trying day, ending well. Began with strong S.W. wind and drift, gradually rising to regular blizzard of about 20 miles an hour. The wind dropped a little, and the surface got better about mid afternoon, and large undulations, a mile or so from crest to crest, began to appear. Mt Nansen hove up, but the view of the mountains is not good owing to thick stratus clouds between us and them. We must have dropped a few hundred feet since lunch I think. Mawson has just given us a cup and a half of splendid thick hoosh, as it is his last cooking night. I take on tomorrow. We feel as if we had come to the down grades at last. 73 miles out.

"Jan 26th

Run 14 miles. Alt. Temp. -10 to zero. Day began with a blizzard but surface fair and down hill. After lunch surface got much better, though sastrugi very large. It calmed and came on such a thick fog as to be like twilight. The mountains were all obscured, there was no sun and no shadow. The wind was uncertain, and puffy. Mawson and professor, without crampons, began to stumble, so we stopped and pitched camp. 59 miles.

"Jan. 27ᵗʰ

Run 16 miles. Alt. 200° by hypsometer at lunch. Temp. zero to +5 or -5. Day began with distant mist, which gradually cleared, giving us a perfect day. We all felt great lassitude, which we attributed to the comparative warmth. Sighted Larsen just after lunch. It is not much fun sighting your objective 50 miles off, you approach it so very slowly. We had a glorious view of Nansen, Baxter and Queensland, with the clouds wreathing away from them. We had one bit of sharp down hill, but on the whole, the going was disappointingly heavy. 43 miles."

Mackay did not write his diary for the next four days.

Tempers must have been very fraught as time was running out for them, the travelling was hard, each of them falling down crevasses. Mawson wrote that his experiences on the thirtieth of January were terrible : "my leg gave excruciating pain for large part of day and have hardly ever had a worse time in my life. Agony all day."

Then the resentment and exasperation with David boiled over. Mawson wrote on the thirty first of January that "Mac called prof a bloody fool once on falling into crevasse, and all sorts of other names." He added that David "is apparently half demented [judging] by his actions – the strain has been too great. He says himself that had he known the magnitude [of the task] he would not have undertaken it".

A little later for the same day he wrote "Mac, it seems, got on to the Prof properly at one halt during afternoon whilst I was reconnoitring. He told the Prof also that he would have to give me written authority [to take over] as commander or he would, as medical man, pronounce him insane."

Mackay's diary continues from the first of February : "7 p.m. Nansen Barrier. We have been going so hard, I have been cooking, and suffering so much from snow blindness for the last few days, that I have had no time to write anything.

"Since last entry, we did two days of 20 miles each, with the sail up, and a blizzard behind us. The going was splendid with a steady down hill, and glassy surface, and we should have done more, but Mawson's knee was strained, my heels badly blistered, and the Prof. hardly able to move."

PENGUINS AND SEALS GALORE

Mackay did not write his diary for the second of February but for the third he wrote : "Noon. Relieved from cooking, thank God. The last week has been so eventful that it would take as much writing about as any ordinary year.

Between the time we decided to come straight down the Larsen glacier and now, we have had so many disappointments and difficulties that I don't think anything can daunt us now. The main point is that we have reached the sea, within sight of our Drygalski depot, but have just been thrown back, by the horn of the little gulf I marked on the map to the north of the depot. This is due to our having approached the depot on a wrong bearing. We must now sledge back round the tip of this horn. But we have penguins and seals galore within sight, and have had our first feed of Emperor [penguins]. The other important event is that I have deposed the Professor. I simply told him that he was no longer fit to lead the party, that the situation was now critical, and that he must officially appoint Mawson leader, or I would declare him, the Professor, physically and mentally unfit. He acted on my proposal at once. We are now, [of] course, expecting the ship. The Professor says that Shackleton promised to send her to look for us on the 1st, but one can't believe a word he says".

Mawson's account of what happened is slightly different. He wrote on the second of February of their pitiful condition : "Prof's boots were frozen on and foot gone. Mac now reports that his feet are more or less gangrenous. During most of the day the Prof has been walking on his ankles. He was no doubt doing his best in this way, and Mac appears to have kicked him several times when in the harness."

Then later he wrote : "The Prof was now certainly partially demented. Yesterday evening, on Mac again threatening him, he [David] had asked me to consider myself leader of the expedition ... as indeed, he said, he had always considered me. He said he would draw it up in writing and get me to sign it. I said I did not like it and would think on it. Whilst Mac was away killing seal he drew out his pocket book and began writing out my authority as leader of expedition and asked me to sign it. I again said I did not like the business and stated he had better leave matters as they were until the ship failed to turn up."

David told the story in a different light – admittedly he was writing months after the event and, as his account was going in Shackleton's book of the expedition, he did not wish to comment on the detail of what actually happened. He wrote [HoA 2 pp203/4] that while Mackay was killing and then skinning the Emperor penguins Mawson was looking for a possible place to camp : "I joined him a few minutes later and as I was feeling much exhausted after the continuous forced marches back from the Magnetic Pole, asked him to take over the leadership of the expedition. I considered that under the circumstances I was justified in taking this step. We had accomplished the work assigned to us by our leader, having reached the Magnetic Pole."

David added that although they only had a very little food at the depot ("two days supply of broken biscuits with a little cheese") there were plenty of seals and penguins which they could kill and eat.

Again he developed, indeed, rationalised, the point about Mawson taking charge.: "even now, in the event of some immediate strenuous action being necessary, if the *Nimrod* were to suddenly appear at some point along the coast, I thought it best for Mawson, who was less physically exhausted than me, to be in charge. He had, throughout the whole journey, shown excellent capacity of leadership, fully justifying the opinion held of him by Lieutenant Shackleton when providing in my instructions that in the event of anything happening to myself Mawson was to assume the leadership. When I spoke to him on the subject, he at first demurred, but finally said that he would act for a time, and would think the matter over at his leisure before definitely deciding to become permanently the leader. I offered to give him authority in writing as leader, but this he declined to receive."

Significantly, David made no mention at all of Mackay' actions or his own mental condition, but again this must have been to try and avoid publication of Mackay's threats to declare him insane and the reactions that such publication would have received. But, equally significantly, David was to conclude his narrative of their journey with no particular mention of Mackay at all or the key role he played in their success.

It was the next day – the fourth of February – when the *Nimrod* did see them and then picked them up : but that was touch and go.

Mawson stated in his diary that Mackay thought they should only wait until the tenth of February before starting home : David and Mawson thought they should wait until the twentieth of February. Mackay was clearly the more realistic because there was no way the *Nimrod* would still be looking for them up to the twentieth of the month. Unbeknown to Mackay the ship's captain – Frederick Pryce Evans, who had commanded the *Koonya* the year before on the journey from New Zealand – was of the same view but even more so: there was no way he could continue searching for the shore party for more than one or two days. Coal was getting short and he had to get back to Cape Royds as soon as he could.

Mackay did not write his diary for the fourth or fifth of February. But on the sixth of February he wrote his last entry for their journey :

"Feb. 6th

8 a.m. While we were inside the tent, having our first good feed of fried seal meat, cooked over the blubber lamp, we heard a gun go off. Mawson jumped up yelling – 'It's a gun from the ship'. This was about 4 p.m. Sure enough it was the ship, come right into the creek, and lying within half a mile of us. We all ran to the water's edge, and Mawson went bang down a crevasse more than 20 feet deep.

David and Mackay tried to lift Mawson up from the crevasse but they did not have the strength. John King Davis, the first officer of the *Nimrod* led a party from the ship to where Mawson had fallen and, placing some timber from the ship across the top of the crevasse, he was lowered down to Mawson. Davis tied the rope around Mawson and he was pulled up and then Davis himself was pulled up.

Just after Mawson had fallen down the crevasse Mackay, rushing towards the ship, had cried out 'Mawson has fallen down a crevasse, and we got to the Magnetic Pole' – though Mackay did not mention this in his diary.

ENOUGH TO MAKE A MAN RELIGIOUS

Mackay's diary went on, first about Mawson in the crevasse :

"This might have been very serious, but luckily there was soft snow at the bottom, and he was not much hurt. I can't possibly tell all that happened next, for I must confess my eyes were a little dim.

In less than no time, we were eating anything we could lay our hands on, drinking bubbly wine, and revelling in the sight of friendly faces and the sound of friendly voices. Almost all the news was good, though there was no news of the southern party. We were relieved from a very real peril of death. I had made up my mind that if the ship did not turn up on the 5th or shortly after, we might pretty well give her up. We would then have started down the coast, with all our rations exhausted, that is to say, nothing to live on but seal meat, and with our tent, and clothes utterly worn out. The professor could not have lived many weeks and his weakness would have delayed us to such an extent as to finish us. The whole thing is enough to make a man religious.

I am away sledging again now, bringing in a depot that I have not mentioned, which we left about 10 miles back on the barrier."

In fact, after a few miles and with heavy snow falling, Mackay felt they would not be able to reach the depot, where they had left some instruments, so they returned to the ship. Mawson then later led a small party from the ship's crew to the depot to pick up the instruments and equipment they had left and successfully returned two days later.

Frederick Evans [HL p107] described the party when they came on board the *Nimrod*: they were "a curious looking little group. Abnormally lean ... they were the colour of mahogany with hands that resembled the talons of a bird of prey."

On the seventh of February the *Nimrod* headed out for Cape Royds and reached it four days later.

Louis Bernacchi had estimated the position of the Magnetic Pole as 72° 51' south, 156° 25' east years earlier. Mawson had estimated its position in January 1909 – as far as his instruments would allow – as 72° 25' south, 156° 16' east. Subsequently it was somewhat ironic that David – in a letter to Mawson [25 May 1925] – was to comment that in view of Mawson's 1911 to 1914 expedition to the region and the further estimations made then "it is now clear that you and I and Mackay were apparently no nearer to the South Magnetic Pole than Shackleton and his party were to the South Geographical Pole."

Shackleton's Southern Party got to within ninety-seven geographical miles of the South Pole. Although the magnetic poles, both north and south, constantly move, it is difficult to see how the South Magnetic Pole could have moved between Shackleton's expedition and Mawson's later expedition to such a large distance and therefore David's point must have been valid: they had fallen short of the South Magnetic Pole by some fifty miles but not as much as ninety seven miles.

Current estimates are that the magnetic poles move up to five miles a year and the latest location of the South Magnetic Pole [2005] are that is located at 64° 53' south and 137° 56' east.

Even so, the journey of David, Mawson and Mackay, given the weather, the ice, the mountains, the crevasses, the awful privations and, above all, the unknown terrain, to where they thought was the location of the South Magnetic Pole, was magnificent. And, although, David was not to refer to the overall contribution of Mackay to the party's success, and Mawson's comment on Mackay being a good soldier but not a good general was to stand – it is clear that Mackay's contribution was crucial.

Awkward, temperamental, irascible he may have been. But he was strong, powerful, had rigged up the sail, made the blubber lamp, caught and killed most of the seals and penguins they had for food, and pulled the sledges consistently. They could not have done without him.

The last wry comment by David on it [HoA 2 p221] was "It is easy, of course, to be wise after the event, but there is no doubt that had we known that there was going to be an abundance of seals all along the coast, and had we had an efficient team of dogs we could have accomplished our journey in probably half the time it actually occupied."

But the three men – in spite of everything – had made it to where they thought the Magnetic Pole was – and they had got back safely.

After they had got back to Cape Royds, and after the whole party had got back to New Zealand, there is no evidence that Mackay ever communicated with David or Mawson again. It was a sad footnote to what had been a magnificent journey.

SOUTH AND NORTH

SOUTH AND NORTH

When the *Nimrod* got back to the expedition's hut at Cape Royds there was no sign of Shackleton's Southern Party. Shackleton had started on his southern journey on 29 October 1908 and his party had carried food for ninety one days. When the *Nimrod* arrived at Cape Royds on 11 February 1909 Shackleton's party had been away for one hundred and six days : what had happened to them?

Shackleton had left instructions with Murray when they had started and had named 28 February as the last date by which they should reach Hut Point (where Scott's *Discovery* expedition hut was) on their return. Shackleton had stated that if he and his party had not returned by 25 February then a relief party of three men was to be landed at Hut Point, together with a team of dogs. If they had not reached them by 28 February then the relief party should start the next day to go south to look for them. If the relief party found no trace of Shackleton's party then they should stay over for the Antarctic winter and look for them again in the spring by which time they would undoubtedly be dead but the relief party might be able to find out what had happened to them.

Captain Evans, the ship's captain, was all the time anxious that they should not get frozen in the ice and therefore should start back for New Zealand at the earliest opportunity. Partly because of this, and also partly because he was a strong willed and somewhat domineering character, he insisted on taking charge of all those at Cape Royds and thereby, in effect, displacing the position of Murray whom Shackleton had left in charge during his absence. Because of this there was considerable disquiet over the arrangements for the relief party and who should pick its members : but Evans insisted that this was his responsibility.

Mackay's diary entries resumed on the fifteenth of February. After the *Nimrod* got to Cape Royds Evans decided it was best for the ship to leave there and lay off Glacier Tongue to avoid being trapped by ice. Although the *Nimrod* was to go between the Tongue and Cape Royds Evans preferred to spend most of the time in the relative shelter provided by the Tongue.

Mackay's first entry was dated the fifteenth of February and covered the period since they reached the ship on the sixth. He mentioned the abortive trip he had tried to make to get to the depot they had left eight miles inland.

"**S.Y. Nimrod Glacier Tongue 15. 2. 09**

I spent one uncomfortable night in the tent with Harbord [the ship's second officer] and two seamen in a three man bag, and then returned to the ship. We found it impossible to keep our course by compass, and as the fog showed no sign of lifting, and we were only two miles from the ship, we thought it as well to spend our time comfortably on board, as in the tent, where everything was wet from thaw. It cleared up a bit next day and Mawson went out and brought in the depot.

"While we had been out, I had read bits of my diary to Harbord, in the men's presence (unavoidable) to pass the time. Soon after I came on board, the skipper called me to his cabin, and spoke to me in a serious, although perfectly friendly way, about this, saying that I had made a rash disclosure to the crew, and that the 'Daily Mail', with which Shackleton has made an arrangement regarding a monopoly of our news, has been making a fuss about some disclosures made by some members of the crew in New Zealand when they had got back there after landing the original shore party..

"The end of it was, I had to give up my old diary, which was put away under seal."

There was nothing unusual in this as on each of the expeditions, at that time, there were strict rules laid down about the publication of diaries and the timing of their publication and if Shackleton's arrangement with the Daily Mail had been jeopardised in any way that might have affected the money that was being paid to him. And Shackleton – both on this and his later expedition – was to be worried continually about money.

When David, Mawson and Mackay had got on board the *Nimrod* they had eaten much food – in fact Mackay had eaten far too much. His diary continued: "The next event of importance was a very severe stomach ache, which attacked me one night. It was the worst I have ever had. Of course it was simply the result of over eating, but it was the worst that I can ever remember. I literally could not move, and sometimes could hardly breathe."

Mackay – in spite of his own medical education – tried to shift the blame for this : "People returning from sledging, should be fed, at any rate for the first three days, on carefully chosen, digestible food, instead of being pressed by well meaning friends to take second helpings of plum pudding, Irish stew &c.

"As soon as Mawson had returned (on the 7th) we left relief inlet, and headed for Cape Royds. The weather had been either foggy or thick with snow and drift all the time we had been in relief inlet. But the wind now got up from the South, and blew a pretty smart blizzard. In addition to this, large floes kept drifting down on us, so we did not get to Cape Royds till the 11th. Here Mawson and I landed, he to stay but I only to get my bedding &c.

"I saw Murray, Priestley and Bobs [William Roberts], but had so little time that I could hardly exchange a word with them.

"The ice was all out of back door bay, and the ice foot pushed farther back than last year, so that there was much more rock exposed along the water line.

"The ship now sailed for Hut Point but was stopped about three miles distant from it by fast sea ice. We have been tied up to the N. side of Glacier Tongue until the 14th. With a mild blizzard blowing. The southern party are now overdue, and we are beginning to feel anxious about them. I am feeling very lazy and rather like a stuffed turkey."

His next entry was four days later.

"Glacier Tongue   19. 2. 09

On the sixteenth we ran out and had a look at Hut Point for signs of the Southern Party. We got within about two miles of it, stopped by fast ice, but saw no signs. I was surprised that Captain Evans did not send a sledge party to the hut, but he didn't, as we had to make the most of the fine weather. So we dropped down to Cape Royds, found all well, landed them two bags of coal, and brought off Murray, his goods and chattels and any delicacies I could find.

"Murray is a great addition to the company, and I am glad to have him aboard."

James Murray had travelled down from New Zealand in the *Nimrod* with Mackay : in fact they were the only two members of the shore party to go in the *Nimrod*. Both were Scottish – Mackay a tall, broad shouldered, Highlander and Murray a short squat Glaswegian. Murray was a biologist with a particular interest in microscopic zoology. He was forty three years old and, apart from Edgworth David, the oldest man on the expedition. Murray was to join Mackay on the ill fated Canadian Arctic Expedition of Vilhjalmur Stefannsson four years later.

Mackay's diary continues :

"For the last four days we have been lying in our snug little harbour at the tip of the Tongue, and it has been blowing a steady blizzard."

"Glacier Tongue   21. 2. 09

The blizzard took off a bit yesterday evening, so we steamed up to Hut Point to have a look for the Southern Party.

"As we approached, about 10 p.m. a figure was seen walking on the sky line, and excitement became intense. Would we find eight men, or only four (the supporting party)? Our doubts were soon put to rest, for only the supporting party, Joyce, Mackintosh, Day and Marston, with eight dogs, had come in.

"We soon had them on board, and jolly glad we were to see each other. They are a cheery crowd, who have evidently been the best of friends all through, and who would be enough to brighten up an evangelist.

"They have done their work most thoroughly, performed some wonderful marches and are loud in their praise of the dogs.

"They report having encountered some very dangerous crevasses, from which they had at least one narrow escape.

"We put them all ashore with Murray, at Cape Royds this morning and took Mawson off. There was a strong breeze from the S.E. and the landing was difficult. We returned to Glacier Tongue."

In his next diary entry – three days later – he mentioned Evans' comments on a proposed relief party and the strong feelings on it. Mackay himself was to be the most vociferous on the composition of the party. It is perhaps surprising that so many of the shore party felt so strongly about staying behind and waiting for Shackleton's party, even if it meant having to spend another winter at Cape Royds. It was perhaps a tribute to the loyalty which Shackleton had engendered among his men.

Mackay wrote :

"Glacier Tongue   22. 2. 09

This morning it was fine, calm and bright, so we steamed up to Hut Point, with a view to surveying the hut, and finding a good berth for the ship; so that we might be able to land stores quickly if it becomes necessary to land a party. The hut was in a very untidy and dirty state, and there were large drifts of snow inside it.

"Captain Evans announced to us this evening that he will inform us exactly of his intentions on the 25th. In the meantime he says that after this date, he will begin preparing to leave a party of six ashore, half at Cape Royds, half at Hut Point. Everybody is very keen to stay, and some people, of whom Brocklehurst is one, will be very much hurt if they don't stay, feeling themselves stellenbosched if they are sent home. Also Brocklehurst declares that he believes someone is lining his pockets with the Expedition funds.

"But there are so many 'sinister rumours' floating around of friction between members of the Southern Party and friction between supporters of the expedition at home, that I don't think they are worth noting."

"Glacier Tongue   23. 2. 09

"Another fine calm day.

"The edge of this glacier against which we are lying has been breaking away in chunks of several tons weight at a time during the days of heavy northerly swell that we have had for the last few days. But this erosion is so dependent on weather conditions that it would be useless to try to state its rapidity in terms of area washed away per day or week or month. But certainly these falls of large masses occurred at least every half hour in a length of barrier edge of about half a mile observed. The fall was all from above water level, and so a shelf is left sticking out below the water line. I never observed this shelf rise to the surface in masses; but a constant stream of bubbles rises from it, showing that it must be melting.

"At the water line the sea wears a slot or groove in the barrier edge, which may extend fifteen feet or so into the ice, before the overhanging shelf is broken off by the swell. This slot is about a foot wide. Under it there is, as I say, a shelf, about three feet thick, and when the surface of the water is smooth another slot may be seen below this and another shelf.

"A most important observation is that, just where we are moored to the glacier, there must be a cave, or tunnel under the ice, permitting the passage of a current of water from one side to the other. This tunnel is not very wide, for slack water may be seen at either side of the stream which issues from it. The stream eddies up from under the glacier, and runs northward at almost a knot an hour. It attains a maximum of strength twice daily. It is probably the chief factor in making this bight, or harbour in which we lie."

It was after this that the row about the composition of the relief party – in which Mackay was a key protagonist – blew up. And there was still no sign of Shackleton's Southern Party.

In his instructions to Murray Shackleton had said that Mawson should lead the relief party but had not stipulated who the others should be except that they should be 'volunteers'.

Mackay wrote :

"Glacier Tongue   24. 2. 09

"A fine day. But we lie here inactive, watching for signs of the Southern Party at Hut Point.

"Captain Evans took me to his cabin today, and had a long talk, but most of it was soft sawder [sic]. He told me definitely that tomorrow he would select the six men for a party to remain behind during the winter, but would give no hint as to who they were to be."

"Glacier Tongue   24. 2. 09

"Another fine day spent in inactivity, absolute inactivity. Except that a watch is kept on Hut Point for signs of the overdue party. We are all in a state of the most painful anxiety. And in addition to this we are all made uncomfortable by the rather hole-and-corner business that is going on.

"A relief party of six men is to be left at Hut Point to spend the winter and wait for the missing people. How these men are to be chosen has not been disclosed. But I fear that a strong effort will be made to make this finishing up of the expedition a colonial business, to the discredit and dishonour of the British members of the land party. Anything of this sort I shall strongly resent, and I have prepared a protest, which I shall present if I think necessary."

David, Mawson and Armytage were all Australians. If they were to be on the relief party then – if Mackay's suspicions were correct – the others, presumably, would be Australians or New Zealanders from the ship's crew. What was then particularly to infuriate Mackay was the rumour that Dr Rupert Mitchell, the Canadian surgeon on the *Nimrod*, was to be the doctor to winter with the party and not him.

What made the atmosphere even more difficult was the fact that Evans made it clear that in his view the task of the relief party was to 'find the bodies' of the members of Shackleton's party. There was even a rumour that Evans was not too concerned if Shackleton's party did not make it back as he relished the idea of being in overall command : but it is difficult to believe that there was any truth in this.

Mackay wrote down his letter of protest to Evans in his diary :
"Sir,
In accordance with permission, which you expressed to me verbally during our conversation on the 24th, I have the honour to submit the following protest to you.

"I do so with sincere respect and regard to you as Captain of the ship.

"It appears to me that Mr Kinsey [the English born shipping magnate from Christchurch in New Zealand who was the expedition's agent and who had appointed Evans as captain of the *Koonya* when the expedition had sailed to the Antarctic and who had been involved – though not decisively – in the decision to appoint Evans, rather than Mackintosh, as captain of the *Nimrod* on the return journey to the Antarctic] has already, on two occasions, taken arbitrary action, in opposition to Lieut. Shackleton's clearly expressed wishes, and that further arbitrary action is contemplated.

"The two actions I refer to are :

1) The supersession of Mr Mackintosh as commander of the ship

2) The practical supersession of Mr Murray as commander of the land party in Mr Shackleton's absence.

"The action which I believe to be contemplated is the landing of a relief party in which the present land party will not be fairly represented.

"This action is arbitrary, inexpedient and unjust in the gravest sense of the word. Arbitrary – because Mr Murray should certainly have a voice in the selection of the relief party, Inexpedient – because men of long experience and proved worth are to be replaced by untried men of no experience.

"<u>Unjust in the gravest sense of the word!</u>" [Mackay had underlined this comment in his diary]. Because (other things being equal) men should be chosen who are most closely bound to the missing members of the expedition by the ties of old friendship and natural affection. This point will become most important if letters, perhaps addressed to individuals, or relics of the missing people are found. It has been urged, that personal feelings must not be considered in the selection of the relief party. But personal feeling of the sort I have mentioned, does exist; it cannot be suppressed, and further, it should do good as an incentive to effort.

"Another point is this. There is no doubt that the reputations of all members of the land party must suffer seriously, should the question be asked – 'Why did you desert your leader, or make no attempt to rescue him?' – or – 'Why did you leave the work of rescue to strangers?' And, in view of what has already appeared in the newspapers at home I cannot help thinking that these are questions which we may almost expect from a certain section of the press and from relatives of the missing people.

"I now come to my last, and least important point.

"If Dr Mitchell is selected for the relief party instead of myself, then I am sent home superseded and disgraced. I am confident that there has been nothing in my whole conduct, while in the service of the expedition, to justify this action. If, in your opinion, there is, I request that it may be made public. No one can deny that I am superior in physique to Dr Mitchell. I have proved myself more than efficient at sledging, while he has not. It is admitted that I have proved my worth on more than one trying occasion.

"Every one of the land party will admit that I have not been stupid, or lacking in zeal in the practical work of the expedition. My medical qualifications are as good as, if not better than, those of Dr Mitchell.

"It may be urged that I have had my share of honours, and that Dr Mitchell should now be given a chance. But this is a weak plea; for the main consideration in selecting the relief party should be to choose men who are most likely to do good work during the search which is contemplated.

"I conclude by assuring you of my continued respect and regard. This letter is written in no insubordinate spirit, but after long consideration.

"I am most anxious to carry out all your orders and wishes in connection with the work of the relief party zealously and cheerfully. But I feel that my first duty is to the gentleman whom Lieut. Shackleton appointed as my commander during his absence, namely, Mr Murray."

Mackay made several good points in his letter and it was clear that he felt very strongly on the issue. It appeared, though, that Evans had consulted Mawson on the matter and Mawson's view would undoubtedly have been based on his experience of Mackay during the South Magnetic Pole journey.

Mackay's diary continued.

"Feb 27th Glacier Tongue

"Blowing hard yesterday and today. We lie here inactive except for getting stores and coal for the land party on deck."

"Feb 28th Glacier Tongue

"Last night I presented my protest to the skipper.

"The result was a long confabulation in whispers, some of which were inaudible.

"The skipper led off by declaring that he and Mawson were determined not to put me ashore with the relief party. This was his irrevocable will. On asking him his reasons for taking this step, which would be so detrimental to my character, he brought forward one insufficient reason after another, among others, that I had grossly offended Shackleton. But, as I pretty satisfactorily disposed of these, he at last told me that there was one reason stronger than all the others, which he would not disclose unless I gave my word of honour to keep it secret. This I could not promise to do off hand.

"This morning he sent for me again, and, in Mawson's presence, informed me that the objection which I have mentioned was this, that several members of the party in the hut at Cape Royds had declared that they believed I was mad.

"I replied that this was simply laughable. Also I said that I thought Mawson was acting in a rather unfriendly way. Mawson then spoke straight out, and said that he had made up his mind to take only 'quiet chaps' on shore with him. And there the matter pretty well ended. I shook hands with both of them."

The situation was partly because of the absence of clear direction on the leadership at Cape Royds. Shackleton had designated Murray to take charge in his absence but had also told him that Mawson should lead the relief party, if one was needed. Shackleton had not spelt out who the other members of the relief party should be other than that they should be 'volunteers'. Then Evans, as ship's captain, had insisted that he was in overall charge.

The composition of the relief party as finally agreed between Evans and Mawson was never listed in detail. Apparently Evans and Mawson did agree to add Priestley and Mackintsoh to their original list of Australians and New Zealanders – and perhaps Joyce – but exactly who was on the final list remains unclear.

In any event, when Shackleton came back on board he immediately assumed command and all dissension disappeared.

During that critical period from the twenty fifth of February until the first of March it was clear that Evans failed to carry out Shackleton's instructions and this could have jeopardised the rescue of the whole Shackleton Southern Party.

Shackleton had made it clear that if he had failed to get back to Hut Point by the twenty-fifth of February then a relief party of three men should be landed there. If his party failed to get back by the twenty-eighth of February then the relief party should set out south to look for them. If the relief party could find no trace of Shackleton's party then they should winter over and look for them in the Antarctic spring – in Evans' words, 'look for the bodies'. The relief party of three would be part of the larger relief party who would spend the second winter there.

Evans, however, did not land the initial relief party of three men at Hut Point. The ship did sail from Cape Royds on the twenty first of February and could fairly easily have landed the party on the twenty fifth. In stead the ship had laid off Glacier Tongue.

Frank Wild summed up the position in his memoirs [ML MSS 2198/2 p54]: "presuming upon Shackleton's death, many of his instructions had been disregarded, including the installing of a lookout on Observation Hill, which would have saved us many hours of mental and physical agony. However, Shackleton forgave those responsible and never made mention of it, so I will let it go at that."

Mackay recounted the details of the rescue when he resumed his diary on the fifth of March.

"S.Y. Nimrod 5. 3. 09   2 a.m.

"The last few days have been very eventful. I may be out in my dates, as we have been living very irregularly.

"I think it must have been on the 1st at about 9 a.m. that we left Glacier Tongue, and proceeded to Hut Point, with the intention of landing a working party to get the hut ready for habitation. I had, somewhat rashly, volunteered for this party.

"As the ship approached Hut Point, some of us were sitting in the Wardroom, when we heard a cheerful yell, and a clatter of feet, on the deck above.

"We all tumbled out, and rushed forward to the foc'sl head. There was a crowd there, some saying they had seen a flash signal, and some a figure beside Vince's cross."

The Southern Party of Shackleton, Wild, Marshall and Adams had been struggling with great difficulty to get back in time to meet the ship before it sailed. Shackleton and Wild had pushed on by themselves on the morning of the twenty seventh of February. Marshall was suffering badly from diarrhoea and they had left him and Adams in the tent. Shackleton and Wild got to Hut Point on the twenty eighth but there was no one there. They saw a note from David pinned to the door of the hut saying that his party had reached the South Magnetic Pole – but that was all. They did not know why the ship was not there, nor a relief party, nor even if the ship was waiting for them at all.

Shackleton and Wild tried to tie a flag to the cross erected in memory of George Vince (who had died on Scott's *Discovery* expedition seven years earlier) but their fingers were so frozen they could not tie the knot. They spent the night wrapped up in some roofing felt they had found. In the morning they did just manage to tie the flag to the cross and when they saw the ship, having left Glacier Tongue to make for Hut Point, come into view they signalled with the heliograph.

Mackay's diary continues : "Soon we could make out two figures plainly, and the excitement was tremendous. We all danced about and cheered and waved our arms, and then fell to punching each other.

"We sent a boat ashore and found that the two people were Shackleton and Wild, both well, and with the news that they had reached 88° 33' S. at a height of 10,000 feet and that Marshall had been left sick with dysentery, with Adams to look after him, twenty five miles out on the barrier."

Shackleton and Wild had a meal of bacon and fried bread on board the *Nimrod and* then Shackleton picked the men to accompany him back to where he had left Marshall and Adams. This time there was no discussion or argument. Shackleton was the leader and he would decide. He chose Mawson, an athletic looking stoker from the ship named Thomas McGillan – and Alister Mackay. They set off three and a half hours after Shackleton had got on board the *Nimrod*.

Shackleton must have been exhausted; he had been travelling for the better part of two days and had had no proper sleep for the previous fifty five hours. But the next couple of days were also a tremendous strain. Mackay wrote : "Shackleton only stopped for lunch, and then started out with Mawson, myself, and a foc'sl hand [McGillan] to bring in the other two. We started from the barrier edge, S.W. of Pram point, did 25 miles out on to the barrier, [reached Marshall, who had recovered his health a bit, and Adams] and on returning Shackleton decided to make for Hut Point, as the ship had not turned up at the barrier edge to meet us as arranged."

Once again, Evans had not carried out Shackleton's instructions. He was desperate not to let the *Nimrod* get frozen in the ice : but his actions nearly wrecked again the rescue of Shackleton's party.

Mackay wrote: "We had to do this [get to Hut Point] over the hills, and it added about 10 miles to our journey, so we did a good deal of travelling, with moderately heavy sledges, in our two days. Coming over the hills, we had to leave everything behind but our sledge and sleeping bags. So we arrived at the hut without a cooker. Here Shackleton's resourcefulness came out; for he soon had an excellent hoosh cooked for us in an old butter tin. We burnt a flare, by simply bursting open a tin of carbide, pump-shipping [urinating] on it, and setting alight to it. It went off with a slight explosion. It was seen by Mackintosh, on board the ship at Glacier Tongue, and in a few hours we were on board.

"Everyone was now on board, but a great quantity of valuable gear, including personal property and geological and biological specimens, had been left at Cape Royds, Hut Point, and at various depots along the coast.

Wild, Shackleton, Adams, Marshall after their return from furthest south © Cantebury Museum.

"But the season was growing late, young ice was forming, there was a good deal of swell, and a breeze from the S. So the end of the business is that at this moment homeward bound, bumping our way through this season's ice, which is in the form of pantiles some three or four inches thick. I have left a great many things behind that I am very sorry to lose."

When Shackleton and the others had left *Nimrod* to fetch Marshall and Adams, Captain Evans had said to Wild [memoirs pp54-55] "Shackleton is a good goer, eh?" When Wild replied "somewhat forcibly in the affirmative" Evans had said "ah well, he has a party there that will see him out." Two days later when they had got back Wild recorded [ibid] "that Mackay fell into the wardroom crying out to the ship's doctor, 'Into they hands, O Doc, I deliver my body and spirit'. He and Mawson went to bed for two days, the all round athletic stoker [McGillan] went to bed for five days while Shackleton went on the bridge and conned the ship out of the bay. What I said to Captain Evans may not be recorded".

Mackay wrote two other pieces in his diary.

"March 9th

There is no doubt that Shackleton's arrival relieved a situation which was becoming very strained.

"Since the fifth, we have had wonderful fine, calm, weather. We have been cruising about twenty miles off the shore in a westerly course from Cape Adare, and have passed to the westward of Cape North, Ross's farthest west, and mapped in some 40 miles of new coast line. The coast is all very mountainous, with some splendid peaks in the background. The hills meet the water in rocky cliffs or very steep snow slopes. It would be almost an impossible country to get over once the sea ice had gone out.

"We have been steaming through the pantile ice. I have spoken of the whole time, and the reason we are not closer in shore is that we are afraid of sticking in it. The sea looks rather like a pond covered with huge water lily leaves.

"Life on board resolves itself into a monotonous round of eating and sleeping. Wild has put on 21 pounds in four days. So he says. We live like fighting cocks or prize pigs."

His last entry was on the twenty first of March :

"We are to sight land (Stewart Island) tomorrow. Nothing of importance since last entry. Weather has been pretty bad. Wind mostly from N.W. Conditions of life on board are almost sickening, for the tables, seats, floor and walls of our eating rooms are plastered with grease, treacle and plum duff, and one can eat nothing of a fluid nature without spilling most of it over oneself or one's neighbour. Under such circumstances we maintain a forced gaiety, which it takes a considerable effort to keep going. We sing a good deal; and I am thinking hard about a scheme for a new expedition. Sometimes we haul on ropes. We have no other amusements."

The *Nimrod* reached Lyttleton in New Zealand on the twenty fifth of March. There the party split up and those returning to Britain went by different ships.

There was an amusing end piece to Mackay's time on the *Nimrod* expedition. It was recounted in the book 'Antarctic Days' written by James Murray and George Marston and published in 1913. The book was a light hearted collection of events and anecdotes from the expedition and the journey of the party from New Zealand to Antarctica. Mackay had contributed an appendix to the book on well known sea shanties. The amusing piece was one of the author's account of the fate of Mackay's 'piece of rock'.

The story went [AD pp174/5] : "Mackay had collected a large rock of granite in the Antarctic as a souvenir. He took great pains of his 'rock' and carefully nursed it back to civilization.

"Whether he risked it on the *Nimrod* or tended it himself through all the vicissitudes of a protracted and much broken journey through New Zealand, Australia, India, France and possibly other countries, I have not been informed. In either case it would travel something approaching 20,000 miles and, by whatever route it travelled, it reached London safely, and was taken charge of by its owner to be conveyed in a cab to his hotel.

"After surviving all the dangers of a sledge journey and of much intricate sea-voyaging, the granite was lost in London, being forgotten in the cab."

Mackay, following his reference in his diary, did draw up tentative plans for his own expedition to Antarctica. In essence he proposed going along the Antarctic coast from Graham Land (the Antarctic peninsula) in the west to South Victoria Land in the east. He and a party of five others would be landed by ship in December (probably in 1911) and then they would go with three sledges and twenty four dogs along the coast (including going along the edge of the Great Ice Barrier), living off the seals and penguins they could catch, kill and eat. Each of the sledges would carry a weight of some four hundred and forty pounds which would in theory have proved fairly easy, but they would be going on foot where no one else had gone. Whether there would be sufficient wild life for them to live off was unknown. What the conditions – in terms of the weather, the coast, the surface and the Ice

Barrier – would be for such a journey was also unknown and the distance involved would have been considerable.

Mackay must have thought of the idea based on his own journey along the coast of South Victoria Land towards to where they turned inland to the South Magnetic Pole. But, apart from the one reference to the detail of this plan by a writer in a German journal [Wichmann in Petermann's Mittelungen 1911], there is no other reference in any other book or journal to it. It was rumoured, however, that Mackay apparently expressed the hope that his proposed journey could somehow be linked to William Speirs Bruce's plans for another Antarctic expedition of his own : this was perfectly possible as Bruce did know Mackay and had particular dealings with James Murray, both before and after the *Nimrod* expedition.

Bruce had led the Scottish National Antarctic Expedition of 1902 to 1904 in the *Scotia*. It was primarily a scientific expedition and Bruce, a strong Scottish nationalist, was proud of the fact that all the expedition members were Scots The expedition was 'to carry on deep sea and other research in the Antarctic Ocean to the south of South America' and 'to carry on systematic observations and researches in meteorology, geology, biology, topography and terrestrial physics'. It was highly successful and much work was done analysing and researching the results and findings when the *Scotia* returned. James Murray was one of those particularly involved in this research work.

They had mapped much of the coast by the Weddell Sea, named part of it Coats Land after the principal sponsor of the expedition and Bruce's party had wintered over in Antarctica. Bruce's reputation was considerable when the expedition returned but he never achieved the public recognition that he should have received – mainly because public interest was focused on the *Discovery* expedition of 1901 to 1904 and particularly the 'furthest south' of 82° 17' south achieved by Scott, Wilson and Shackleton.

It was ironic that Bruce's expedition was so well regarded within the relatively closed world of scientists and polar enthusiasts but not in the outside world. He himself had originally hoped to join Scott's *Discovery* expedition. At one point (16 March 1900) Sir Clements Markham, then the driving forcer behind Scott's expedition and the leading figure of the Royal Geographical Society in London, had written to Professor J W Gregory, the first appointment as scientific director of the *Discovery* expedition (who was later to resign and be replaced by Dr Edward Wilson), suggesting Bruce as his second in command. Gregory had objected and the matter was dropped. Subsequent hopes by Bruce to be on the expedition were dashed (and it would appear that Markham did nothing further to press for Bruce to be included in Scott's party) and it was therefore somewhat gratifying later for Bruce to be able to tell Markham that he had subsequently arranged his own expedition to the Antarctic. Markham was considerably displeased and called the Bruce expedition 'a mischievous interference' [AGEJ letter to WLMcK 22May1978] ; as far as he was concerned, the Scott expedition was the major Antarctic event and anything else a distraction.

Bruce got to know Shackleton well when he returned from the *Scotia* expedition and especially when Shackleton was, for twelve months from 1904 to 1905, appointed as the secretary/treasurer of the Royal Scottish Geographical Society based in Edinburgh. As stated previously, it was probably Bruce who had suggested James Murray and Alister Mackay to Shackleton as members of the *Nimrod* shore party in the Antarctic.

Bruce then led several surveys and small expeditions in the following years to Spitsbergen, the major island in the Svalbard group of islands to the north of Norway in latitudes 77° to 81° north. Murray, as a trained biologist, helped to do research work on some of the materials and specimens that Bruce brought back. From some of the letters that Bruce and Murray exchanged – particularly in March 1911 – it is clear that Murray was hoping that he might go with Bruce on a further expedition to Spitsbergen. But by then Bruce was thinking seriously of something much grander: the second Scottish National Antarctic Expedition.

Bruce's plans were developed in 1910 and published early in 1911. A ship would land ten to twelve people hopefully somewhere along Coats Land in the Weddell Sea and from there a party would attempt to cross Antarctica via the South Pole and meet up with a supporting party coming to meet them – probably by the Beardmore Glacier – from where the ship had dropped them in McMurdo Sound.

In any event there were several others thinking of leading their own Antarctic expeditions at that time and most came to nothing. Some were just sketchy ideas – like Mackay's; others were fairly detailed proposals – like Bruce's.

Frank Wild had been thinking at one time of making his own attempt at getting to the South Pole in 1911.[FWp133]. Early in 1910 Edward (Teddy) Evans, who had served on the relief ship *Morning* to Scott's *Discovery* expedition, had thought of a similar expedition to Mackay's but abandoned this idea when he joined Scott's *Terra Nova* expedition as the second in command [McK/AGEJ 2Nov1976]

There had also been a proposal in the United States for Robert Peary, who had claimed to have reached the North Pole in April 1909, and Bob

Bartlett, who had been the ship's captain on the Peary expedition and who had gone as far as one hundred and thirty miles from the North Pole with the last of Peary's supporting parties, to lead an American Antarctic Expedition with the idea of crossing the Antarctic continent via the South Pole. In the light of the controversy in the United States about whether Dr Frederick Cook or Peary had actually been the first to reach the North Pole, the idea of such an Antarctic expedition was dropped.

There were other attempts – some successful – to launch Antarctic expeditions at this time. A Japanese expedition, under Lieutenant Nobu Shirrase left Tokyo in December 1910 to land near the Bay of Whales (where Roald Amundsen, the Norwegian explorer did make his base the following year and from which he successfully reached the South Pole in December 1911) but it only managed to land two small parties, one which went inland a short way and the other which was the first to visit the Alexandra mountains in King Edward VII Land.

There was the abortive scheme of an Englishman, J Foster Stackhouse, to sail in Scott's old ship the *Discovery* and land a party to go from Graham Land along the coast to the Bay of Whales This was similar to the idea that Mackay had. But, in the end, Stackhouse could not raise the money for his idea and it was abandoned.

Mawson towards the end of 1909 was proposing to go to the Antarctic again and, having failed to agree Captain Scott's terms for joining his *Terra Nova* expedition, proposed to go from Cape Adare along the coast in east Antarctica for some two thousand miles in what was then called the Australian Quadrant, landing three different parties along the way – although in the event only two shore parties were landed.

There was also talk of an Anglo-Swedish expedition being launched under the leadership of Otto Nordenskjold (who had wintered on an island off Graham Land – the Antarctic peninsula – during the Swedish South Polar Expedition of 1901-1904) and then an Austro-Hungarian expedition under Felix Koenig. A particular project by the German explorer Wilhelm Filchner for a possible trans-Antarctic crossing was also, at one time, rumoured to involve Bruce's own plans.

Scott's ship *Terra Nova* had reached Melbourne in Australia on the way to Antarctica for Scott's second expedition at the time Bruce was still trying to raise money for his plans. Shackleton was also rumoured to be planning a further expedition. There was little public notice of or interest in what Bruce was trying to do, either from Scots or anyone else. Although his published plans concluded with the stirring words 'The total cost of the expedition will be about £50,000, and its departure depends on the enthusiasm and patriotism of Scots at home and abroad' he never got any significant publicity or money for them to come to fruition.

When Bruce was drawing up his proposals it is clear that he had Murray in mind to go with him as biologist and oceanographer and had asked him to sound out Mackay. In one of his letters Murray said he had not been able to contact Mackay at all but then in a subsequent letter [20Mar1911] said he had managed to locate him and that he was abroad and would back in Britain in May of that year. By then Bruce's plans had been virtually abandoned and so Mackay's ideas – of either going on his original journey or as part of Bruce's own party – also came to nothing.

But, apart from his rather vague idea of another Antarctic expedition, what was Mackay now thinking of doing once he had returned from the *Nimrod* expedition? He had distinguished himself by being one of the party to be the first men to reach the summit of Mount Erebus. He had been one of the three who had reached what they had determined was the South Magnetic Pole. He clearly had been well thought of by Shackleton when he included him in the small party to go back to Marshall and Adams from where Shackleton and Wild had left them on their return journey from 88° 23' south.

He had left the Royal Navy, having served in it for four years, before the *Nimrod* expedition. He was now thirty years old, a qualified doctor with military experience in the second Boer War of 1899 to 1902, and unmarried. He was a strong and healthy man. He may have been short tempered and what Dr Eric Marshall had called 'a little eccentric' but he had energy and ambition : so what was he to do?

When back in Britain, Mackay wrote in July 1909 to George Marston (known as 'Putty') in response to a suggestion by Marston that they went on a walking tour, He stated that he was staying at his married sister's house in Lanarkshire. The letter [19 July 1909] was rather sad in that he wrote "I can't tell you anything definite about my plans yet, except this, that after paying off my most pressing debts, I shall have little more than £100 left, and I must economise, so I can't think of going for any long walking tour just now. In fact I would very much like to lay my fingers on that there bonus" [HRO].

In fact there never was any bonus actually proposed for members of the expedition though some members were to claim later that Shackleton had never delivered on some promises of future money.

Mackay ended his letter : "Give my chin chin to the others. You need not tell Shacks of my financial straits."

After some time searching for a job he secured a position as ship's doctor with the shipping company owned by David and William Henderson – D & W Henderson of the Clyde – a Scottish company which both built and owned ships and ran many voyages to the Middle and Far East and also to New Zealand and Australia. They were mainly trading ships but some carried emigrants to New Zealand : they were not cruise liners. There is no record of how many different ships Mackay served on but it is known that during his voyages he did visit Rangoon in Burma, Bombay in India and Colombo in Ceylon.

During the *Nimrod* expedition Mackay was not the most popular member of the shore party but he had tried to mix with them all. Murray had recorded that during the Antarctic winter Mackay, Marston and Wild 'frequently enlivened an evening for us with shanties and folk songs and other things. If anybody were not in a mood for music he was free to go and bury himself. The general good came first.' [ADp152]

But Shackleton barely mentioned him in his book on the expedition (The Heart of the Antarctic) and Frank Wild – a man who could get on with most people – did not mention him at all in his memoirs. Mackay, though, did get on with Murray and did keep in intermittent touch with him when they both returned to Scotland. He had also been friendly with Marston and exchanged correspondence with him but, significantly, there is no record of him having any contact at all with either Edgworth David, Douglas Mawson or Shackleton himself.

It was after Mackay returned to Scotland that he started to drink heavily. There is no record at all of his having drunk to excess during the *Nimrod* expedition (where alcoholic drink was scarce anyway) and while he would have had the opportunity to drink a fair amount during his days as a Royal Navy surgeon when the ships docked in various ports and countries, it was only after his return from Antarctica that his drinking became a serious problem, and presumably his journeys with the Henderson Line did allow access to much more alcohol than had been available on the *Nimrod*.

Late in 1912 Marston telegraphed him at his old address in Poplar in east London – presumably inviting him to his wedding – and Mackay replied [19 December 1912] : "I got your wire, forwarded from Poplar. Thanks for invite. I heard wonderful news of you in London, and was very anxious to meet you, but they kept me bobbing about so much, looking for ships, that I really had no time.

"Please receive my congratulations on your happy state, and let me have a long letter about you. They tell me you are starting a poultry farm."

Mackay ended his letter with : "I am going into a Home for Inebriates, for a year!!!"

The 'Home' was Lathallan House in Colinsburgh in Fife. Marston did write back to him with a long letter and Mackay in turn replied. There is no date on this letter but it must have been early in 1913.

Mackay wrote : "It was good of you to write me such a long letter. As to your suggestion of 'pluck' displayed in coming to this place, I cannot see where it comes in. The life here would be a very pleasant one, but for the feeling of wasting time and money, and attaching an additional stigma to my already rather (or very) dubious reputation.

"My doctor in Edinburgh attributed my frequent lapses from the paths of strict sobriety to the kick on the eye which I received in S.A. [South Africa] I told him that I thought this a very charitable view to take of the case."

Mackay then wrote about the house in which he was staying. It was set in nine hundred acres of land and had some forty bedrooms with twenty people staying in it. His letter reads rather pathetically in that he describes his days spent in little activity and not much interchange with the others whom he described as "inebriates or narcomaniacs. Many of these are old habitués, who come here at irregular intervals, to avoid or recover from their excesses. Their conversation, though sometimes amusing, is not very improving, consisting as it does principally in descriptions of their various busts".

Mackay finished his letter by writing "I will always be glad to hear of you." But there is no record of any further correspondence between Mackay and Marston.

It was early in 1913 that the Canadian explorer Vilhjalmur Stefansson visited Britain and met several times with William Speirs Bruce. He had plans for a major scientific Arctic expedition. It was Bruce who suggested Murray be part of that expedition and – though there is no documentary evidence for this – it appears highly likely that Murray also suggested that Mackay should be part of it. If Murray knew of Mackay's stay in the 'Home for Inebriates' (as he almost certainly did) it is quite possible that Murray suggested this as a way for Mackay to break free from his alcoholism.

However it precisely happened, Mackay joined Stefansson's Canadian Arctic Expedition in the spring of 1913.

THE CANADIAN
ARCTIC EXPEDITION

THE CANADIAN ARCTIC EXPEDITION

Although Mackay might have thought that his joining the Canadian Arctic Expedition of Vilhjalmur Stefansson would have helped him overcome his dependency on alcohol and give him a new purpose in life, it was not to be. His participation in that expedition was to lead to his death at the young age of thirty-five.

Stefansson's parents were from Iceland. They had emigrated to Manitoba in Canada to take up farming. Stefansson was born there in 1879 but, shortly after, the family moved to North Dakota in the USA and brought another farm. It was there that he grew up, went to school and then graduated from the State University of North Dakota. He later studied anthropology at Harvard University.

Between 1905 and 1912 Stefansson had travelled extensively in northern Canada, Greenland and Alaska in different expeditions, serving as an ethnologist and an anthropologist. He made studies of the various groups of Eskimos he both travelled and stayed with. In Victoria Island, one of the larger islands to the north of the Canadian mainland he claimed to have discovered a group of 'Blond Eskimos' whom some fanciful accounts (but not Stefansson himself) at the time credited with being descendants of some survivors of the expedition led by Sir John Franklin.

That expedition, under the authority of the Royal Navy, was meant to find the sea route from the Atlantic to the Pacific Oceans over the top of Canada – the fabled north west passage. It had sailed from Britain in 1845. It resulted in the death of Franklin, the one hundred and twenty-nine members of the expedition and the loss of his two ships, the *Erebus* and the *Terror*. Different theories as to the fate of all the men and the reasons for their death are still the subject of much debate today.

Stefansson was not a particularly physically impressive man – he was only of medium height – but he was to build a considerable reputation as an expert on the Eskimo way of life and how to travel and survive in the Arctic. He wrote several books : his most famous was 'The Friendly Arctic' (first published in 1921) wherein he claimed that it was wrong to regard the Arctic as a sterile and barren environment and that it was possible to live and travel by living off the natural wild life that abounded there.

In 1912 after returning from his expedition with the Canadian zoologist Dr Rudolph Anderson he now proposed an expedition – with Anderson as his second in command – to explore the area of the Beaufort Sea between Alaska and northern Canada and the North Pole. This was regarded as the largest unexplored area in the world. It was called by some the 'Zone of comparative inaccessibility' because no one had ever been there and it was extremely difficult to access. There was much speculation that perhaps there was a large area of land – even a continent – somewhere there between Canada, Alaska and the Pole.

Stefansson was to write in the Canadian *Victoria Times* [June 1913] "The sensational aspect of the Canadian Arctic Expedition is that if it should prove as successful as it conceivably may be, then it will close forever the chapter of geographical discovery, for the only place on the whole earth where there can possibly be land of a conceivable extent whose very existence is unknown to us, is the unexplored area of a million or so square miles that is represented by white patches on the map, lying between Alaska and the North Pole".

Stefansson wanted to be the man who explored that major area previously unexplored, particularly if a new continent was to be discovered. In fact he did not go far out on to the ice during his 1913 to 1918 expedition, though he did spend considerable time around the islands north of the Canadian mainland, and so the large unexplored area remained just that. Ernest Shackleton was later in 1920 to have a similar idea for exploring for land in the Beaufort Sea but abandoned that to go south to Antarctica again in his ship *Quest*. Thereafter major explorations of the area would be made by air.

Stefansson had secured promises of 45,000 US dollars for the expedition from the American Museum of Natural History and the National Geographic Society but it was not enough. He approached the Canadian Government and they offered to fund it if they could take over the whole expedition. Stefansson – without consulting anyone – agreed and thus it became the Canadian Arctic Expedition.

When the National Geographic Society conceded that the Canadian Government would take over authority – and funding – for the expedition it made it clear that the expedition must begin in May or June in 1913 : otherwise the Society would reaffirm its original authority and would launch the expedition the following year. Whether they would have been able to do this is questionable; but the result was that Stefansson was now short of time. The Canadian Government insisted he should start in summer of 1913 and that, as well as Stefansson's own somewhat arbitrary and unstructured

decisions, meant there never was any coherent selection of equipment or men for the expedition.

The first thing was to get a ship. Stefansson purchased a twenty nine year old wooden whaling ship, the *Karluk*, for ten thousand dollars and, although cheap, the purchase was his first big mistake. Although it underwent many overhauls and repairs in the short time available, the ship was totally unsatisfactory at the start of the expedition. Bob Bartlett, whom Stefansson had asked (although his second choice) to be Master of the ship, told one of the Government naval ministers that in his view the *Karluk* would never be able to make the voyage and was "absolutely unsuitable to remain in winter ice" [Diu.p71]. In spite of his considerable reservations, however, Bartlett agreed to captain the ship on its journey. He later [LogBBp256] stated that the voyage was "the most tragic and ill fated cruise in my whole career".

Then the Canadian Government decreed that the aims of the expedition should be expanded and that another ship would be needed. Stefansson bought the *Alaska*, to be used mainly as a supply vessel but also as the ship for a southern party. This would be under the command of Rudolph Anderson and would conduct studies around Coronation Gulf and the islands off the northern coast of Canada. The northern party, under Stefansson on the *Karluk*, would go further north, look for new land and conduct scientific studies : as Stefansson rather grandly put it, the party would – as well as search for possible land – "investigate tides, currents, the depth and character of the Arctic sea bed, the temperature, chemical composition, and vegetable and animal life of the North Pacific and Arctic Oceans" [Kp5].

Then, later, Stefansson bought a third ship – also to be used as a supply ship – the *Mary Sachs* (and much later bought a fourth ship, *North Star*). Because of the tight timetable involved ,some of the equipment for the different parties was on the wrong ships and some of the men earmarked for the two separate parties – northern and southern – were also on different ships. At the beginning Stefansson thought that James Murray should captain the *Mary Sachs* but when that ship sailed Murray was on board the *Karluk*. To bring order into what was fast becoming chaos, Stefansson said that the three ships (*Karluk*, *Alaska* and *Mary Sachs*) would all rendezvous at Herschel Island, north of the Alaskan coast at 72° north, and sort out the correct equipment and men there. As it turned out, none of the three ships actually got to Herschel Island.

The selection of the crew was also unsatisfactory. Bartlett arrived at Victoria in Vancouver Island after the crew had been chosen, and only days before the ship was to sail (he had only been asked to be the Master of the *Karluk* at the

Mackay with four of the ponies brought from Siberia

beginning of June). He soon dismissed the first officer for incompetence and in his stead appointed the twenty two year old Scot, Sandy Anderson, who had signed on as the second officer (who was not related to Anderson, the second in command of the expedition).

According to William Laird McKinlay [Kp12] one of the crew was a confirmed drug addict "who carried around a pocket sized case with half a dozen phials of drugs and two hypodermic syringes' (the ship's cook, Robert Templeman) while 'another suffered from venereal disease". And ,although there was not meant to be any drinking of alcohol on the ship, the chief engineer (John Munro) did manage to smuggle a large amount of alcohol on board. Mackay was able to get hold of much of that, partly because as he claimed, as the ship's doctor, that he used the whisky for medicinal purposes.

Bartlett, in his own account of the voyage, seemed reluctant to criticize Stefansson, particularly as apparently it was Robert Peary who had recommended him to Stefansson and Bartlett always stayed loyal to Peary. But he did later write about the men of the expedition [LogBB257] and having mentioned them by name stated "I must point out that some of the above men had never seen sea ice before. I do this because, without the slightest idea of criticism, I want to show how this very lack of experience ultimately led to the tragic death of a number of our party".

McKinlay was a twenty-four old school teacher in Glasgow. He had graduated from Glasgow University in mathematics and natural philosophy and had worked for some years on the magnetic and meteorological observations brought back by William Bruce from the *Scotia* expedition. It was Bruce who had recommended him to Stefansson. In April 1913 he received a telegram from Stefansson inviting him to join the expedition as magnetician and meteorologist. He immediately accepted but later somewhat wryly wrote [Kp10] that he was told by a member of the Canadian Meteorological Service that he had been hired – at a monthly salary from the Service of sixty Canadian dollars (nearly twice his salary as a teacher) – because no one in the Meteorological Service "was willing to be associated with Stefansson's expedition".

Over sixty years later McKinlay's account of the expedition and what happened to those on board the *Karluk* was published. It was a devastating critique of Stefansson and his leadership of the expedition.

James Murray was recruited as biologist for the expedition early in 1913 after a strong recommendation from William Bruce to Stefansson, (and sailed with his wife to Vancouver to join the *Karluk)* and Mackay joined as the expedition's doctor shortly afterwards. With McKinlay they were the three Scots in the scientific staff of the party. Although Stefansson had claimed he would "employ only British subjects wherever British subjects are available" [MS1M1913] it was only the three Scotsmen who came from Britain. Of the other eleven scientists (apart from Stefansson himself) five were Canadian, two American, one was Australian, one was from New Zealand, one was Danish and the other was from France.

Stefansson also claimed that his men would form "a larger staff of scientific specialists than have ever been carried on a Polar expedition" [Kp5]. But the lack of polar experience among most of them was to prove crucial – even for Murray and Mackay.

At the start of the expedition Murray was forty-six years old, still somewhat squat but robust and in good health. He seemed to get on with most of the expedition party. Mackay was thirty-four years old, six foot tall and still muscular but increasingly quick tempered, impatient and unwilling to tolerate anything he deemed as inefficient or unsatisfactory. He seemed to get on with few others. Some of those who survived the journey from the shipwreck of the *Karluk* to Wrangel island and then the appalling conditions on that island before they were eventually rescued, were critical of both Murray and Mackay. Notable among these was Ernest Chafe, the young assistant steward among the crew, who subsequently wrote his account of the expedition [MMBC]. McKinlay, though, had nothing but praise for Murray.

In correspondence with AGE Jones, the English polar historian, many years later [24Oct1976] McKinlay conceded that both of them were critical of Bartlett and were anxious to break away from the rest of the party once the *Karluk* had sunk, but said he "had the greatest admiration for James Murray. He was the only one of us who was able to carry out a full programme of work and he was unremitting in his efforts to take advantage of every opportunity to add to the oceanographical knowledge of the area we were covering." McKinlay went to say "He was fully dedicated to his job. I am happy to be able to record that I was closely associated with him in all he undertook and, apart from our difference on policy, our relations were cordial".

On Mackay, however, McKinlay wrote a different story : "I regret to say that he should never have been associated with the expedition or, for that matter, any expedition. He was over attached to liquor and this was from the outset, and continued to be throughout, a major handicap to his usefulness for the differences with skipper Bob [Bartlett] and who was the controlling influence in the plans of the breakaway party."

Given Mackay's attitudes to some of the members of the *Nimrod* expedition it is not surprising that now, a few years older and seriously affected by alcohol, he should have been critical of the *Karluk* expedition members. It was, after all, very badly organised and led. With hindsight, of course, Mackay never should have gone on the expedition – though neither should anyone else. It was not only the state of the ship, the inexperience among the crew and the vagueness of the expedition's purposes, but also the cavalier and inadequate leadership shown by Stefansson once they had started, all factors which contributed to the expedition's failure. And, above all, Stefansson was not Shackleton.

The *Karluk* sailed from Esquimalt, just by Victoria, the capital of Vancouver Island, with Bob Bartlett in charge, on the seventeenth of June 1913. Stefansson, Anderson and James Murray(and his wife) sailed separately in a mail steamer and joined them a few days after the *Karluk had* reached Nome in Alaska on the eighth of July.

Because of what Stefansson described as the "unexpected increase of cargo at Nome" [TFA] it was then that he decided to buy a third ship, the *Mary Sachs*. It was also then that dissatisfaction among the crew and the scientific staff with the arrangements for the expedition surfaced. A number of meetings with Stefansson were held with Murray, apparently, being the most aggressive in raising complaints and worries about food supply, fresh water, travel, equipment and, especially, the lack of clear plans for both the proposed northern and southern parties of the expedition. McKinlay wrote in his diary that "The position of the northern party is serious and no one would be surprised should Murray resign" [Kp15].

On the thirteenth of July the *Karluk* sailed from Nome to Port Clarence, further up the coast of Alaska. There the *Mary Sachs* also arrived and an Eskimo woman (who was to act as seamstress and sew clothes for the expedition) came on board the *Karluk*, together with her two young daughters aged eight and three. There were already two Eskimo men on board and Stefansson was to bring on two more later. Apart from them, and a cat, there were thirteen of the crew and, including Stefansson, ten members of the scientific staff. Although thirteen scientists had been originally chosen, the other three travelled on the other ship.

At Port Clarence Murray's wife left: she was not to see her husband again. The *Karluk* left on the twenty sixth of July. Five days later they met the sea ice.

The ice was about one month earlier than usual and Bartlett, seeing a vast frozen sea ahead of them, was all for turning back. He was not happy at all with the strength of the *Karluk* in the ice and was anxious not to get frozen in – particularly so early in the voyage. But Stefansson insisted they went on, as far as they could (though later, Stefansson indicated that he was unhappy at Bartlett's decision to press ahead).

By the fourth of August the ship was locked in the ice with the wind blowing them on to the Alaskan coast. They were then some twenty five miles away from Point Barrow on the northern coast of Alaska. Stefansson decided he and two Eskimos were go across the ice by sledge to a small trading settlement, Cape Smythe, just near Point Barrow. He also chose Mackay to go with him.

By midnight the Eskimos had returned with the dogs and the sledge. They said that the journey over the rough and broken ice had been difficult, Mackay had fallen in the water and the Eskimos said he was almost exhausted before they reached the land ice. Stefansson and Mackay did go on to the settlement at Cape Smythe, bought some furs, kayaks and two small boats made from animal skins, and returned to the *Karluk* two days later. It was Mackay's baptism at travelling over the Arctic sea ice.

The same day the grip of the ice loosened and Bartlett was able to get the ship moving under steam. By the ninth of August the *Karluk* was working eastward towards the Alaskan/Canadian border and only about two days sailing from Herschel Island where the three ships were to meet and sort out the different parties, the equipment and the food. But the ice had now thickened and they could not get there : they were frozen in again.

The scientists and members of the crew of the *Karluk* were getting most unhappy and much criticism was made of Stefansson. This was compounded by the presence of the Eskimos on board ship – and the reluctance of some of the crew to live near them – and also by a fifty-five year old addition to the crew named John Hadley whom Stefansson had brought on board from Cape Smythe. Stefansson apparently thought he now needed someone with Arctic experience and Hadley was that man and he kept on telling the others this. He had been employed by the Cape Smythe Trading Company and had spent many years in the Arctic; but he failed to get on with the others on board the *Karluk*. Stefansson later claimed [TFA] that he thought Hadley could be transferred to either the *Mary Sachs* or the *Alaska* when they got to Herschel Island, but, of course, they never got there.

One of the effects of the whole of Stefansson's expedition being split into different ships was that his original ideas for the composition of the northern and southern parties could not be realised. While on the *Karluk* Stefansson had decided that both James Murray and William McKinlay would go with the southern party on the *Alaska* – most of their equipment was already on that ship. As it was they had travelled on the *Karluk* because there was more room for them there – and, as Stefansson kept telling everyone, it would all be sorted out when they got to Herschel island.

By the twelfth of August the *Karluk* was firmly stuck in the ice. Three days later they were temporarily stuck aground; the ship soon got free but then was stuck in the ice once more. It was now clear that the *Karluk* could not go eastwards any more and the ship started to drift with the ice.

Before the ship drifted too far away Stefansson decided that a party should try and go over the ice to get a message through to a small police outpost on Herschel Island and also call at the nearby Flaxman Island and there leave messages for the captains of the *Mary Sachs* and the *Alaska*. The party set off on the twenty-ninth of August but the surface of the ice was very difficult and it was clear that they would not be able to get far at all. The attempt was called off and the party returned to the ship.

Thereafter the *Karluk* drifted in the ice, at first some twenty miles a day and then thirty, and later, up to sixty miles a day. For the first two weeks they drifted in a circle above the northern coast of Alaska but then by the third week of September they were being carried westwards and any hope of getting to Herschel Island now was abandoned. Equally, it was very difficult to see how the proposed northern party could now get organised.

At first, although there was considerable disquiet at their plight, and criticism of Stefansson for not being clear as to what they should now do, the atmosphere on board was tolerable. The scientists and the crew had separate mess rooms and they each had a gramophone. Mackay would insist on playing a song by Harry Lauder over and over again at first to the delight and then the exasperation of the others . Some of the scientists played bridge. Mackay suggested boxing contests and took part in several.

By the middle of September Stefansson knew that there was no hope of any progress towards the aims of the expedition and he then decided on an extraordinary course of action : he would leave the *Karluk* and not return.

On the twentieth of September Stefansson handed Bartlett a letter. In it he said he would be leaving the *Karluk* and heading for the Jones Islands just north of the Alaskan coast. From there he would hope to make for either Point Barrow or possibly Herschel Island. He would hunt for caribou to supplement their meat supply. It was a strange statement because he had previously told Bartlett that caribou were almost extinct in the area [B HHp5] - and, anyway, how could he expect to get back to the *Karluk* if the ship drifted away to the west? There was no immediate shortage of meat and the Eskimos on board were adept at hunting and killing seals. Stefansson concluded in his letter that "It is likely that we should be back to the ship in ten days, if no accident happens".

Stefansson took with him Burt McConnell, the twenty-four year old secretary to the expedition, George H Wilkins the photographer (who, as Hubert Wilkins, was later to travel over both large regions of the Arctic and Antarctic by air), Diamond Jenness the anthropologist and two of the Eskimos, together with a sledge and twelve of the dogs. It was a strange party to take for hunting caribou.

Stefansson must have realised that staying on board the *Karluk* would achieve nothing and if he was make some success out of what was rapidly becoming an expedition disaster, he should leave the ship. It was a strange thing, though, for an expedition leader to do.

In his letter Stefansson had also said that Bartlett should try and erect "four lines of beacons, running in the four cardinal directions from the ship to as great a distance as practicable. There should be some arrangement by which these beacons indicate in what direction the ship is from each of them. And some of them should have the distance of the ship marked upon them." He further said that the beacons should be erected on "accessible islands". The idea was ludicrous.

Curiously, though, Bartlett stated [Log BB p 260] "I was much impressed by this letter. It certainly was in detail and it showed how thoroughly Stefansson thought things over before he acted" but he also added "If I had any misgivings it was because I had been acquainted with the Arctic for a good many years. You can make all the plans you want to ... but the finer plan you have the worse it will go to smash when wind and ice and drifting snow take charge. That's exactly what happened to Stefansson's plan."

But Bartlett knew that Stefansson did have considerable experience and had spent several years travelling in the Arctic and must have realised the plans set out in his letter were almost incapable of realisation. In fact, two days after Stefansson had left, a forty-mile-an-hour wind blew and the *Karluk* was carried away, stuck in the ice and moving westwards at about thirty miles a day.

Stefansson's party did reach one of the islands just four miles north of the Alaskan coast and from there went to Point Barrow. He learnt that the *Alaska* and the *Mary Sachs* were moored off Collinson Point, further along the coast. It was there that he found the men who were to form the southern party of the expedition (though some of them were stuck on the *Karluk*). The remainder of the southern party, under Rudolph Anderson, did carry out much scientific research for the next two years. Stefansson was to travel in and around the islands, the ice pack and the Canadian mainland for the next four years : but the original objectives of the northern party, – to look for large areas of land, and possibly a new continent (which had been one of the main points of the Canadian Arctic Expedition) – were never pursued.

In the meantime the *Karluk* continued to drift at the mercy of the ice and the wind.

The day before Stefansson left the *Karluk* he had asked two men from the ship to go over the ice to see if they could spy any land. At that time they were some thirty four miles away from the Alaskan coast. Bjaerne Mamen, the young Norwegian topographer, and Alister Forbes Mackay were the two: they went to the west but two days later returned. They had not seen any land.

That journey – proving to Mackay that they could move across the ice, in spite of the difficulties encountered in his earlier trip with Stefansson – and Stefansson's leaving the ship the day before they returned, must have convinced Mackay that the expedition was doomed but that he would be able to travel to some sort of safety once they were near land.

Mackay, and several of the others, were all aware of what had happened to the *Jeanette* over thirty years before.

In July 1879 an American expedition under the command of Lieutenant George De Long had sailed in the *Jeanette*, a barque rigged coal burning ship, from San Fransisco and up to the Bering Strait – the stretch of water between the eastern side of Siberia and the west coast of Alaska. De Long's aim was to enter the ice north of the Bering Strait and hopefully break through it into what many believed was an open polar sea and from there sail to the North Pole. The idea of an ice free Arctic Ocean had many supporters then and De Long was going to exploit it in order to get to the North Pole.

The *Jeanette* got frozen in the ice just east of Wrangel Island and for twenty one months drifted in the ice north of Siberia. In June 1881 the ship sank and most of the ship's party, including De Long, subsequently lost their lives in trying to get to the coast of Siberia and from there to safety.

Both Mackay and Murray discussed the fate of De Long and the *Jeanette* many times in the evenings with the rest of the scientists and the crew. They all wished to avoid the fate of De Long's expedition but it seemed as if Bartlett had no idea how to do this. Undoubtedly Mackay and Murray would have contrasted the leadership of Bartlett with their own experience of Shackleton's leadership on the *Nimrod* expedition : and the contrast was not in favour of Bartlett.

For the rest of September and during October Bartlett did hope that the ice around the *Karluk* would break up and allow them to sail under their own power. But it was not to be. Freezing cold – earlier than in previous years – together with strong winds meant the *Karluk* was locked in the ice. All on the *Karluk* were also worried whether the ship could withstand the strong pressures that the ice brought on the ship. From October the *Karluk* leaked and was letting in water and the pumps had to be used daily : the key question was whether the leaks could be controlled so that the ship would not sink.

Bartlett wrote in his autobiography [Log BBp263] that by the twenty-eighth of October "I gave up hope of getting out of our mess that fall" though he did keep a faint hope that the *Karluk* would survive and hopefully manage to break free of the ice in the spring of 1914.

On the twelfth of November the sun went below the horizon and the long winter night enveloped the ship and its crew. By mid-November the *Karluk* reached its furthest north - 73°N. Then Bartlett ordered a large amount of food, stores, coal, wood and equipment to be off loaded on to the ice in case anything did happen to the ship.

Mackay had secured his own separate cabin on the ship, next to Bartlett's. The others shared accommodation. Many times, in his cabin, Mackay and Murray would discuss their precarious position and how they might be better off by leaving the ship and the others and make their own way to safety. Henri Beauchat, the thirty-four year old French anthropologist, joined them in the cabin. Bartlett could hear them as they discussed their plans and, indeed, the three of them seemed to make no secret of their plans. This knowledge was compounded when both Murray and Mackay started making harnesses for man-hauling sledges, the method of travel used on Shackleton's expedition but, as they were to discover, not much use over the treacherous broken ice of the Arctic Ocean.

Beauchat, the expedition's anthropologist, was an odd character for a polar expedition. He was physically not strong and had no experience at all of polar travel, or indeed, any form of rough travel. He was a weak man and might well have been in some awe of the reputations of Mackay and Murray from their Antarctic experience.

At one point Mackay asked Mamen to join them in their separate party but he steadfastly refused. Bartlett was dismissive of their ideas but did say that he would not stop them from going and would give them some of the

stores and equipment they would need – but, after that, they would be on their own.

The Arctic winter in that December seemed colder than ever. Sometimes the temperature dropped to minus 32° Fahrenheit. The *Karluk* was firmly stuck in the ice. No one who was not on the *Karluk* had any idea of where they were. There was no radio or any other means of contact with the outside world. What had happened to De Long and the *Jeanette* continued to haunt the minds of the ship's party.

The whole party on board the *Karluk* had never been one cohesive unit and the long winter night, with little if any hope of getting home, caused inevitable dissension : Mackay and Murray were outspoken in their criticisms of the expedition, of Stefansson for leaving them and of what they viewed as the lack of leadership by Bartlett. They now made little attempt to conceal their decision to leave the rest of the party and make their own way to land. [K p55].

There were, however, attempts by Bartlett to raise the morale. On Christmas day he authorised a large dinner for them all. On New Year's Day he arranged games, a three legged race and even a shot putt competition – and an attempt at a game of football on the ice – and Mackay organised a chess tournament.

Mackay and Mamen, the Norwegian, reached the final of the chess tournament. Mamen won and received a gift of fifty cigars which Bartlett had promised to the winner.

By the end of December Bartlett estimated that they were some fifty-five miles away from Herald Island, a small and barren outcrop, which he estimated was itself some thirty eight miles east from Wrangel Island. He could see that if they were to reach Siberia then they would have to use Wrangel Island as a base but how were they to get there? The majority of the party were totally unused to sledging over the Arctic Ocean and none had much experience of handling dogs.

On the first of January a strong easterly gale blew and the pressure of the ice on the ship's hull increased. For the next ten days the pressure increased, the wind continued to blow savagely and the gloom of the party intensified. Then on the tenth of January the pressure of the ice finally caused part of the hull by the ship's engine room to crack and water flooded in. The *Karluk* started listing over.

Bartlett gave the order to abandon ship. They moved as much of the food and equipment as they could on to the ice near the ship : clothing, pemmican, sugar, milk, biscuits, coal, gasoline, wood stoves, timber the remaining dogs and nine sledges. Bartlett called the place on the ice where all the stores and equipment were placed 'Shipwreck Camp'.

Pencil sketch by George Marston of Mackay and Wild playing chess at Cape Royds 1908. © R.G.S.

Bartlett remained on the ship until the next day. There were the two gramophones on board with one hundred and fifty records. On one of them was recorded Chopin's Funeral March. As the ship listed more and more, then in the afternoon Bartlett played the Funeral March. As it was playing he left the ship and joined the others at Shipwreck Camp and the *Karluk* sank.

The ship had sunk where the *Jeanette* had got firmly stuck in the ice some thirty five years earlier. Bartlett knew that, if they stayed where they now were, the drift of the ice would carry them westward "and circle the Pole until some years hence, we would possibly emerge down through the Greenland Sea" [Log p268].

In fact this was what had happened to pieces of the *Jeanette*: the movement of the currents and the consequent drift of the ice over the Arctic Ocean had carried wooden remains of the *Jeanette* to Greenland where they were then discovered. That was one of the key factors that influenced Fridtjof Nansen, the celebrated Norwegian explorer, who had already made the first crossing of Greenland in 1888, to develop his plans for getting his ship (the specially constructed *Fram*) stuck in the ice north of Siberia in the hope that the subsequent drift would carry them over the North Pole.

On the twenty first of January Bartlett sent four men to reach Wrangel island with three sledges and eighteen of the dogs. Hopefully they could establish some sort of base there for the main party to follow and then go from there to Siberia. The four were the first mate, the twenty two year old Scot, Sandy Anderson, the second mate Charles Baker, and two seamen – John Brady and Archie King (who for some reason had signed on with the original ship's party under the name of Edmund Lawrence Golightly).

There was later a suggestion that Bartlett had intended to send them to reach Herald Island first and to use that as a base to go to Wrangel Island. Bartlett himself, however, stated that the party set off from the beginning for Wrangel Island [Log BBp270]. As it was, the four-man party did get to Herald island but thought they were approaching Wrangel Island.

Bartlett also sent Mamen and two of the Eskimos as a support party for Anderson and his three companions. Travel for all of them over the treacherous surface of the Arctic Ocean was difficult. Anderson's party got to within three miles of land but were held up by clear water between them and the shore. They then realised that it was not Wrangel but Herald Island : a large lump of rock under five miles long and one mile wide. Wrangel Island was much bigger and mountainous and nearly forty miles away to the west. Anderson reckoned that now they had got so close to Herald Island they had to land there and then try later to cross to Wrangel Island. Mamen and his two Eskimos decided they could do no more as a support party and turned back for Shipwreck Camp. Anderson's decision was a fatal one. He and his three companions were never seen alive again.

In October 1924 a ship called at Herald Island and the skeletons of the four men were discovered along with the remains of their tent and some equipment. They had indeed got to Herald Island but no further. Curiously, after their bones had been taken back to Canada and examined it was decided that they had not died of starvation or disease but, more probably, from kidney failure caused by eating too much pemmican (and not enough fresh seal meat) while at the same time suffering from hypothermia. There was also a brief theory at that time that the bones found were those of Mackay, Murray and the two others who had gone with them, but this, after the examination, was soon discounted.

Bartlett still persisted with his idea of sending out small parties from Shipwreck Camp to lay depots and flags to point the way to Wrangel Island. He realised that that was their only hope of rescuing the party. If they could get to Wrangel Island then a smaller party could try from there to cross to Siberia and get help. Even if the drifting ice and wind destroyed some of the flags and carried some food depots away, there should still be enough of a trail laid for the main party to follow from Shipwreck Camp.

The fact that Anderson had set off for Wrangel Island strengthened the determination of Mackay and Murray to try and make their own way there. At the end of January Mackay and Murray approached Bartlett and said that they, Beauchat and a seaman named Stanley Morris (who had, at a late stage of their preparations, asked to go with them) were intending to leave the camp and make their own way to land : to Wrangel Island and, hopefully, from there to Siberia. Bartlett tried to dissuade them but failed. Murray asked for supplies to last the four men for fifty days and a sledge. Bartlett agreed and also offered for them to take some of the dogs. Mackay declined the offer of the dogs. They would man-haul the sledge.

Bartlett was handed a letter from the four man party on the first of February [LogBB pp271-2]:

"Canadian Arctic Expedition

Sunday Feb. 1ˢᵗ, 1914
Captain Robert Bartlett
  Sir, We, the undersigned, in consideration of the present critical situation, desire to make an attempt to reach the land. We ask you to assist us by issuing

to us from the general stores all necessary sledging and camping provisions and equipment for the proposed journey as per separate requisition already handed to you. On the understanding that you do so and continue as heretofore to supply us with our proportionate share of our provisions while we remain in camp, and in the event of our finding it necessary to return to the camp, we declare that we undertake the journey on our own initiative and absolve you from any responsibility whatever in the matter.

   A Forbes Mackay, James Murray, H Beauchat, S S Morris"

Their supplies included a tent, sleeping bags, spare clothing, a wood stove, fuel, a rifle and one hundred rounds of ammunition. The food they carried seemed more than sufficient to last them fifty days. Altogether the load of supplies and equipment on the sledge weighed some six hundred pounds. It was too much for them.

 The four men set off on the morning of the fifth of February. That afternoon they passed Chafe and Hugh 'Clam' Williams, one of the seamen, making their way back to Shipwreck Camp after placing flags for the subsequent depot parties to follow. Chafe noted that, already, Mackay and Murray had discovered that it was too difficult to drag all their load on the one sledge so they had resorted to relaying : taking half the load forward and then returning to take the other half. It was what Mackay , Mawson and David had done on the first part of their journey to the South Magnetic Pole – it meant going three miles for every one mile forward.

 Ten days later Chafe and two Eskimos – after another depot laying party – were on their way back to Shipwreck Camp when they again met Mackay, Murray and Morris. They were now some twenty miles from Herald Island. This time they were in a pitiable condition. They had tried to lighten their load on the sledge by discarding clothes and equipment as they went. Half of their provisions had been soaked by leaving them on thin ice while they tried to sleep. Their sleeping bags were also soaking wet. Morris had cut his hand while trying to open a tin of pemmican. Their clothes were frozen stiff and their boots practically threadbare. The three men looked exhausted.

 Chafe suggested they return with him to Shipwreck Camp but Mackay and Murray refused. Morris looked too weak to go anywhere. They did accept some seal meat from Chafe but turned down his offer of pemmican. It was clear that Mackay and Murray thought they could still make it to land by themselves and they now said they were going to make direct for Wrangel Island. It was a mixture of pride and not wishing to admit defeat that made them attempt to go on. Chafe thought they would not go far at all and so told Bartlett when he returned to Shipwreck Camp.

After another mile Chafe and Williams came across Beauchat. He had half of the remaining stores with him – presumably waiting for the others to come back and relay them again. He looked worn out and seemed not to be able to walk at all. His boots were practically disintegrating, his hands were swollen and he wasn't wearing his gloves and, according to Chafe, he seemed delirious and suffering from hypothermia. Again Chafe tried to persuade him to go back with him to Shipwreck Camp but Beauchat ignored him.

Chafe summed up their position by telling Bartlett later that he didn't think they would ever see Mackay's party again. "Mackay, Murray and Morris were pulling the sledge by hand. Beauchat was a mile behind with hands and feet frozen and half delirious from his suffering. Morris had blood poisoning from a knife cut in his hand. I think Beauchat must have died that same night." [Log BBpp273-4]

No one saw any trace of Mackay, Murray, Morris or Beauchat again.

AFTERMATH

AFTERMATH

There was nothing Bartlett could do about it. Mackay, Murray and the other two had gone ahead with their plan in spite of his pleading with them not to. From what Chafe had told him, they had also refused to come back to Shipwreck Camp when they must have known it would be suicidal to try and go further by themselves.

In his autobiography Bartlett wrote [Log p274] "I couldn't discuss the situation with Chafe. There was nothing we could do. It would have been a wild goose chase to try to reach the party. Certainly we never could have found them and gotten back. Moreover, a heavy gale was making. I knew the four men were probably dead by now. But I said nothing about it."

With no news from Anderson and his three companions who had been seen last by Mamen three miles from Herald Island, Bartlett realised there were now eight men from the ship either lost or dead. He was desperate to get news of Anderson but it proved impossible for any party to get to Herald Island.

On the nineteenth of February Bartlett and the remainder of the party set off from Shipwreck Camp to make for Wrangel Island. Their journey was difficult in the extreme but they eventually got there three weeks later. Then on the eighteenth of March Bartlett set out with one Eskimo and seven dogs to get to the coast of Siberia. It was just over one hundred miles away and they reached it on the fourth of April. Bartlett estimated that because of the drift of the ice they had in fact travelled some two hundred miles to Siberia.

From there Bartlett went eastwards to the coast of the Bering Strait, helped by some of the Siberian Eskimos he met, – a total of about six hundred miles – and managed to secure passage on a ship (the *Herman*) to the Alaskan port of Nome. It was ironic that the captain of the *Herman* had been Stefansson's first choice for captaining the *Karluk* but he had backed down. It was then that Stefansson had approached Bartlett.

Bartlett finally got to Nome in Alaska on the *Herman* at the end of May. It then took him six weeks to get a ship that would take him to Wrangel Island to rescue his party. The ship was the *Bear*, an American revenue cutter that patrolled that part of the Arctic seas each summer. When eventually the *Bear* was free to try for Wrangel Island it was late in August. The ship got to within twenty miles of Wrangel Island but could get no further because of the ice and had to turn back to Nome.

It all bore a resemblance to the later difficulties Shackleton had in getting a ship to go to Elephant Island to rescue the twenty two men of his *Endurance* expedition that were stranded there.

In Nome Bartlett persuaded the captain of a small trading ship *King and Winge* to call at Wrangel Island on her way to Siberia. Bartlett then joined the *Bear* again for another attempt. The *King and Winge* reached Wrangel Island on the seventh of September and took the survivors from there on board. The next day the *Bear* approached and the survivors were transferred to it. To Bartlett's great distress he was told by them that three of the men on Wrangel Island were now dead – Bjarne Mamen, the young Norwegian, George Stewart Malloch, the thirty-three year old geologist and George Breddy, one of the ship's stokers. Mamen and Malloch had died as a result of the appalling privations the party had endured : bitter cold, lack of food, illness, dissension and especially no idea of when or if they would ever be rescued. Breddy had died from a gunshot wound : whether it was suicide, an accident or a deliberate killing by one of the others has never been satisfactorily determined.

The party were on Wrangel Island for almost six months before they were rescued. Their suffering was intense and what was perhaps the cruellest blow of all was that Stefansson seemed to have shown little interest in them but instead appeared subsequently to put the blame for their plight on Bartlett. From the accounts written afterwards by some of the survivors it appears that many of them had given up any hope of eventual rescue.

It is interesting, though, to contrast their plight with those of Shackleton's men on Elephant Island in the Antarctic in 1916 for four and a half months before rescue. But the men on Elephant Island had Frank Wild to lead them, and they all had faith that Shackleton would eventually come for them.

Altogether eleven members of the expedition on board the *Karluk* had died. There was nothing to show for that part of Stefansson's expedition at all. Even Murray's oceanographic notes were lost when the *Karluk* sank.

McKinlay, one of the survivors from Wrangel Island, joined the British Army in October 1915 and fought on the Western Front with the 51st Highland Division. In his book on the expedition, 'Karluk', he summed up his experience [K p161]:

"Not all the horrors of the Western Front, not the rubble of Arras, nor the hell of Ypres, nor all the mud of Flanders leading to Passchendale, could blot out the memories of that year in the Arctic. The loyalty, the comradeship, the esprit de corps of my fellow officers and of the men it was my privilege to command, enabled us to survive the horrors of the war, and I realised that this was what had been entirely missing up north; it was the lack of real comradeship that had left the scars, not the physical rigours and hazards of the ice pack, nor the deprivations on Wrangel Island".

Before McKinlay was judged fit enough to join the Army he was in correspondence with Mackay's mother several times between April and August of 1915. She had found it difficult to accept that her son, Alister, had died. She wrote several letters to the Canadian Government but never received any reply. She also wrote to the wife of Beauchat but again received no response.

Mrs Forbes Mackay had lost two of her sons, killed in the first year of the First World War. The lack of news of Alister must have been heart breaking. She told McKinlay that one of her sons, then in India, could not accept that he was dead [letter 19Apr1915]. Poignantly she asked McKinlay that "if the Canadian Government does do anything in the way of sending out a search expedition, I hope you will let me know the result". She also wrote, somewhat pathetically,[21 July 1915] that she would like to meet McKinlay (though she never did) as "I want so much to know how you got home in safety because, evidently what you did, is what Alister should have done & might still have been alive."

If Mackay, Murray and their two companions had not tried to make it by themselves to Wrangel Island but gone with Bartlett and the rest of the party, then they might have survived. It is extremely doubtful whether Beauchat would have survived: he was physically weak and should never have gone on the expedition in the first place. Morris could have survived – he was the youngest and strongest of the four of them.

Even so, Mackay, with his body debilitated by several years of heavy drinking, his impatience and quick temper, might easily have succumbed on the long wait at Wrangel Island. It was a tragedy that he died, just days before his thirty-sixth birthday.

His mother finally accepted that he was not coming back from the Arctic in her last letter to McKinlay. McKinlay wrote about this [letter to AGEJ24 Oct 1976] : "When I came home, I received many letters from Alastair's [sic] mother, and the correspondence stretched over a lengthy period. It was clear that she was aware of her son's weakness. I would not have told her, but there was no need for me to tell her. In her last letter she sorrowfully confessed that perhaps it was all for the best that he "was taken, in the hope of a better life in 'the beyond' ".

McKinlay had added "All things considered, perhaps one should be content to remember only the best of him. He did, I believe, make a fine journey to the South Magnetic Pole".

For that, Mackay will always be remembered among the polar men of the 'heroic age' of polar exploration.

NIMROD 1907-1909

Ernest Shackleton   leader

Edgworth David   chief scientist and geologist
Jameson Adams   meteorologist
Philip Brocklehurst   assistant geologist and surveyor
Bernard Day   motor engineer
Ernest Joyce   i/c dogs, sledges and equipment
Alister Forbes Mackay   surgeon and biologist
Douglas Mawson   physicist
Bertram Armytage   i/c ponies
Eric Marshall   surgeon and cartographer
George Marston   artist
James Murray   biologist
Raymond Priestley   geologist
William Roberts   cook
Frank Wild   i/c stores

Aeneas Mackintosh was due to be one of the shore party but the loss of his right eye on 31 January 1908 meant he returned with *Nimrod*.

KARLUK

Canadian Arctic Expedition 1913-1914

Crew :
Robert Bartlett	Master
Alex (Sandy) Anderson	First officer
Charles Barker	Second officer
John Munro	Chief engineer
Robert Willamson	Second engineer
George Breddy	Fireman
John Brady	Seaman
Edmund Golightly (alias Archie King)	Seaman
Stanley Morris	Seaman
Hugh (Clam) Williams	Seaman
Fred Maurer	Fireman
Robert Templeton	Cook and steward
Ernest Chafe	Assistant steward
John Hadley	Carpenter

Scientific Staff :
Vilhjalmur Stefansson	Master
Henri Beauchat	Anthropologist
Diamond Jenness	Anthropologist
Alister Forbes Mackay	Surgeon
George Malloch	Geologist
Bjarne Mamen	Assistant topographer
Burt McConnell	Secretary
James Murray	Oceanographer and biologist
William Laird McKinlay	Magnetician and meteorologist
George Wilkins	Photographer

Eskimos :
Payola 'Jerry'	Hunter
Asecaq 'Jimmy'	Hunter
Kataktovik	Hunter
Kuraluk	Hunter
Kiruk 'Auntie'	Seamstress
Helen	Age eight
Mugpi	Age three

REFERENCE NOTES :

AGEJ	AGE Jones, the late polar historian
AD	Antarctic Days by James Murray & George Marston
B HH	Bartlett – The Great Canadian Explorer by Harold Horwood
BC	Royal British Columbia Museum, Victoria
Diu	Stefansson & the Canadian Arctic by Richard Diubaldo
FW	Frank Wild by Leif Mills
HofA	The Heart of the Antarctic – Being the story of the British Antarctic Expedition 1907 - 1909 by EH Shackleton, 2 volumes
HL	High Latitude by JK Davis
IM	The Ice Master by Jennifer Niven
K	Karluk – The great untold story of Arctic exploration By William Laird McKinlay
LogBB	The Log of Bob Bartlett by Captain Robert A Bartlett
ML	Mitchell Library, Sydney
MMBC	Maritime Museum British Columbia
OMRS	Orders and Medals Research Society Journal, Summer 1997, Vol36 No 2 (235)
TFA	The Friendly Arctic by Vilhjalmur Stefansson

* Mawson's diaries are taken from 'Mawson's Antarctic Diaries' edited by Fred Jacka & Eleanor Jacka, Allen & Unwin 1988
* Mackay's South Magnetic Pole diaries were published by the Royal Scottish Museum, Edinburgh, March 1982, and are quoted by kind permission of the Trustees

Mackay's diaries for latter part of *Nimrod* expedition are held in the Scott Polar Research Institute, Cambridge (SPRI ms1456/7)

* Letters between Mrs Forbes Mackay and McKinlay are in Royal Geographical Society (RGS/LMS/M 8)
* Letters from Mackay to Marston are in the Hampshire Records Office, Winchester, AOO/A75 & 76/1.2.3)
* Letters between AGE Jones and McKinlay are held by the National Library of Scotland, Edinburgh and are quoted by kind permission of the Trustees

PART II
MEARES
1877 - 1937

Terra Nova held up in the pack ice Dec. 1910.

INTRODUCTION

On the morning of the eleventh of December 1911 two men – one an Englishman of Scottish descent and born in Ireland, the other a Russian who could hardly speak any English – turned their dogs and two sledges round to face north.

They were at latitude 83° 30' south, three hundred and ninety miles in a straight line from the geographical South Pole and had over four hundred miles to get back to the expedition hut at Cape Evans by the Ross Sea in Antarctica. They had gone further south than originally intended but, because of the success of the dogs in travelling over the Ice Barrier and the death of the last of the ponies, Captain Scott had taken them some way up the Beardmore Glacier before sending them back. The two men now faced a longer journey home with less food than planned. It would be a very close call; but they made it.

The Englishman was Cecil Meares, the dog driver with Scott's second and last expedition to the Antarctic. The Russian was Dimitri Gerov whom Meares had persuaded to go south when selecting the dogs and ponies for the expedition in Siberia.

The day before, the Norwegian explorer Roald Amundsen had left camp (and last depot) at 88° 23' south – the furthest south reached by Ernest Shackleton and his three companions in January 1909. Amundsen and his four companions were now travelling further south than any other men had ever been before. They reached the South Pole four days later. Scott reached the Pole thirty-four days after them.

Meares left Antarctica two months later when the *Terra Nova* returned to New Zealand. He was never to visit Antarctica again.

His life had been one of travel and adventure in many parts of the world. But it was his involvement in Scott's fateful journey to the South Pole – from which Scott and four others never returned – that his name is remembered.

Meares and Dimitri ready to start South. Nov 1911.

BACKGROUND

Cecil Henry Meares was born on the fourteenth of February 1877 at his grandfather's house in Inistoge, county Kilkenny, in Ireland. His mother died five days later.

Meares' father was a regular Army officer stationed at Curragh in Ireland when Meares was born and then served for several years in India with the Royal Scots Fusiliers. He married again and, when he and his wife were in India, his son's early life was spent with his grandfather, the sisters of his step-mother or in a boarding school.

When Meares was three years old he left his grandfather's house and was taken to Ayr in Scotland to stay with his step-aunts. His wife was later to write a brief chronology of his life [MrsM : BC ms0455] and stated that "he taught himself to read at four without reading lessons – was a very interesting child – but with a very strong will".

At the age of five he went to a local school and then to Ayr Academy, where he boarded with one of the masters. The following year his father and step-mother returned from India and he went to stay with them for a few weeks. His father, an Army major, was then attached to the Bedfordshire regiment and was posted back to India.

At the age of ten Meares went to the Ley's School near Cambridge and then the following year to school in Woking in Surrey, where he stayed for four years. His father and step-mother had returned to England, his father being posted to the Horse Guards in London, and he went to the Woking school as a day boy.

At the age of sixteen Cecil applied to join the Army and, though he passed the initial interview and examination, he failed his medical : his chest measurement was apparently too small. This was surprising as he was to become a tall broad shouldered man within a few years.

Having moved several times in his early life and been to several schools it was then – in 1894 – that he started his travels abroad. His family was fairly wealthy and he did not have to search for a living as soon as he left school. His first journey was to Bilbao in Spain where he stayed and learnt Spanish. The following year he went to Italy and learnt Italian. Then in 1896 he left Europe and sailed for India.

Meares tried his hand at coffee planting and stayed in the southern province of Coorg, in what is now the Indian state of Kerala, for some two years, While there he learnt Hindustani and joined the Coorg & Mysore Rifles, a part time military unit of volunteers, chiefly composed of men from the coffee plantations. The regiment had been formed in 1884 and lasted as a separate regiment until 1917. In 1899 he gave up coffee planting and went to Vladivostok in the far east of Russia and then from there to Peking in China. The same year his step-mother died.

At the end of 1901 Meares went to South Africa and it was there that he joined the Scottish Horse.

His travels up to then had been extensive and it is difficult now to see what exactly he did when visiting the various countries. He did do coffee planting in India but there is no record of anything particular that he did in Russia or China – though he did manage to learn some Russian and Chinese – but travel and dabble in fur trading.

The second Boer War started in October 1899. The peace talks that had ended the first Boer War of 1880-1881, and the London Convention agreement that followed three years later, had given a measure of independence to the two Boer 'republics' of Transvaal and the Orange Free State. But the Boer and the British interpretations of that independence varied. The discovery of gold in Witwatersrand in 1886 increased the British desire – particularly pushed by Cecil Rhodes – for more control over the Transvaal. It seemed that war was inevitable to settle clearly the status of the Boer states in relation to the British Crown and in October 1899 the Boers invaded the British province of Natal.

At first the Boer soldiers greatly outnumbered the British forces but the tide turned when more and more British troops were sent out from Britain and many thousands also joined the British side from different countries of the British Empire.

The Scottish Horse, a cavalry unit, was originally formed in 1900 and recruited from Scotsmen living in South Africa on the initiative of the Caledonian Society of Johannesburg. It had similar terms of service to the other units of the Imperial Yeomanry that were raised to support Britain in the war. Then a second regiment of the Scottish Horse was formed the following year from Scots from home and Australians of Scottish descent.

Meares joined the first regiment and the records show that he landed in South Africa on the twenty-third of December 1901 and was enlisted on that day. It is unclear whether he came straight from India to South Africa or whether he had gone back to Edinburgh where his father was now living and sailed from there. In any event the regimental records show that his profession was listed as 'Planter'. He joined, as others did, for twelve months service or for the duration of the war – whichever was the longest,

Meares joined the regiment (the first Edinburgh squadron) as a trooper and in February 1902 was promoted to lance corporal. The Scottish Horse served with distinction during the Boer War but the only major military engagement that involved them between January 1902 and the end of the war on the thirty-first of May that year was the Battle of Rooiwal in April. A number of individuals in that battle are mentioned in the records but not Meares.

After the Boers surrendered and the Treaty of Vereeniging was signed a number of the Scottish Horse wanted to go back to Edinburgh. Meares was listed as one of those.

In June of 1903 Meares then travelled to Kamchatka in Siberia in northern Russia where he did some trading in furs and learnt the art of sledging with dogs. In September the following year he visited Seattle, in Washington State, in the USA. Then in 1904 the war between Russia and Japan broke out.

EASTERN ADVENTURES

EASTERN ADVENTURES

There has always been a mystery about the involvement of Meares in the Russo-Japanese War of 1904 - 1905.

Herbert Ponting, the photographer on Scott's *Terra Nova* expedition, first met Meares in November 1905 when going on a steamer ship from Yokohama in Japan to Shanghai in China. Ponting wrote [TGWS p4] that "As he had been having a roughish time during the Russo-Japanese War, and needed a holiday, we had come to an arrangement by which he came along with me to act as interpreter and otherwise to assist me in my photographic work". In the ensuing six months Ponting and Meares travelled to India, Burma and Ceylon.

Ponting's reference to a "roughish time" is one of the only two specific references to Meares' experiences during the Russo-Japanese War. The other reference was by Captain Oates.

Oates, with whom Meares became particularly friendly during the *Terra Nova* expedition, recorded [SPRI Mss1317/1.2] of Meares that "he was a prisoner among the Russians in the retreat from Mukden".

The Battle of Mukden was fought in February and March of 1905. The Russian Army had retreated in the latter part of 1904 and adopted a defensive line some fifty miles south of Mukden in Manchuria. Then on the twentieth of February the Japanese Fifth Army attacked the Russian left flank. The main attack on the centre of the Russian line began seven days later. Then the Japanese Fourth Army attacked the Russian line on its right flank.

Fearing they would be encircled the Russians retreated and while at first fighting hard, their retreat soon became a rout. The city of Mukden was evacuated in March. Some fifty-three thousand Russian troops were either killed or wounded and forty thousand taken prisoner. The Japanese lost forty-one thousand men killed or wounded.

Many writers on Scott's expedition – and many of the biographies of some of Scott's men – have stated that Meares was an 'observer' during the Russo-Japanese War. What is not clear is whether Meares was an observer on the Russian or the Japanese side or whether he was taken by the Russians as a prisoner because they suspected him of spying for the Japanese.

Russia had established diplomatic relations with Japan in a series of treaties in 1855 and 1858. At the time Japan was only just emerging from its position of being almost a feudal state and was starting to make contact with the outside world. But the growing friendship between Russia and China caused the Japanese to view Russia with increasing hostility. Britain, also, was for many years suspicious of expanding Russian influence – in her case, particularly in central Asia and Afghanistan.

In 1894 and 1895 the rivalry between China and Japan erupted into war. The Japanese won decisively. Russia then had to choose between getting closer and more friendly with Japan or becoming a sort of protector of China. The Russian Tsar decided to side with China.

The Russians then made a formal alliance with China in 1896 and this established ownership by Russia of the Chinese Eastern Railway which was aimed at making a rail link between Siberia, through Manchuria, to Vladivostok. Two years later they persuaded the Chinese to grant them control of the Liaotung Peninsula which the Russian Government – together with those of France and Germany – had previously persuaded the Japanese to withdraw from. On the coast of that peninsula the Russians built a naval base in the ice free waters of Port Arthur.

The increasing encroachment of Russia caused further anxiety among the Japanese. This was compounded when in 1900 several of the European nations – including Britain – sent armed forces to protect their Chinese embassies from the Boxer rebellion. Russia took the opportunity to move troops into Manchuria. Even at that stage the Japanese might have been willing to concede authority over Manchuria but they wanted in return that Russia should recognise Japan's interests over Korea. This the Russians were not prepared to do.

The increasing closeness of relations between Russia and China increased British suspicions of Russian intentions, especially over Britain's commercial interests in parts of China. The result was that in January 1902 Britain made a formal alliance with Japan.

Intermittent discussions between Russia and Japan continued but made no progress. Then in February 1904 – without warning – Japanese forces launched an attack on the Russian fleet in Port Arthur. The Russo-Japanese War began.

The war attracted considerable interest from the European countries and from the United States of America. Over one hundred journalists from different countries were accredited to the Japanese Government and the Japanese armies in the field at the beginning of the war. One of those accredited as a war photographer to the Japanese First Army in Manchuria, for Harper's Weekly and an American agency, was Herbert Ponting. Another was the American writer Jack London who was correspondent for the San Fransisco Examiner.

The reason for the international interest was clear. The war was the first between a leading European power (Russia) and an oriental country (Japan) and in the end was to result in the first major defeat of a European country by an eastern army since the days of Genghis Khan.

As the war progressed, the Japanese captured Port Arthur in December 1904 after a long siege, then captured Mukden and then defeated the Russian fleet in May 1905 at the battle of Tsushima. The Russian far eastern(Pacific) fleet had already been seriously weakened when Port Arthur fell. It was the Russian Baltic fleet that fought and lost at Tsushima – having sailed all the way from the north west coast of Russia via the North Sea, the Atlantic Ocean, round the Cape of Good Hope and, through the Indian and Pacific Oceans.

It was that defeat that led the Russians to agree to the pressure from President Theodore Roosevelt of the USA for peace negotiations with Japan. These culminated in the Treaty of Portsmouth (New Hampshire) in September 1905 and the formal end of the war.

Japan had thus gained the position of a world power, being the first non-European or non-American imperialist modern state. Significantly Britain was the foremost supporter of Japan in the war and this was reflected much later when the British Air Mission – including Cecil Meares – was sent to Japan in 1921. British attitudes were also typified by some of the accounts of the war written afterwards. One of the most laudatory references to the Japanese was perhaps in General Sir Ian Hamilton's "A Staff Officer's Handbook". Another was in the dedication of journalist Frederick Palmer's book "With Kuroki in Manchuria" : "To the Japanese infantry smiling, brave, tireless and no less to the daring gunners who dragged their guns close to the enemy's line overnight, this book written by one who was with you for five months in the field is admiringly dedicated".

Both Hamilton's and Palmer's books refer to some of the English correspondents as having been present at the Boer War though in neither are any specifically mentioned.

The aspects of the Russo-Japanese War – long sieges, trench warfare, naval battles (the first major battles since Trafalgar in 1805) – were all noted by the British as well as the other major countries and were to give useful lessons for the various countries in the First World War : though not all the countries were to learn from those lessons.

Most of the journalists sent to cover the war were attached to the Japanese side and some undoubtedly also acted as spies for the Japanese. Several Governments also had military attaches formally posted with the Japanese armies and there were also a number of 'observers', a term used to describe an official military or government representative where the observer's own country was not directly involved.

Meares is not listed anywhere as a journalist or war correspondent during the war. If he had been such, then he would have undoubtedly described himself as a correspondent and not an observer. He was not serving in the British Army or the Royal Navy at the time so would not have been any sort of formal military attache. All the members of Scott's *Terra Nova* expedition who referred to Meares in their written diaries or letters all described him as an 'observer' during the Russo-Japanese War and this must have been because Meares described himself as such.

This has all led to speculation that Meares was in some way working for Britain in an intelligence capacity – a speculation that has constantly been referred to in any reference to Meares by writers and polar historians and also in the obituaries of Meares that were published after his death in 1937.

For years – indeed, for centuries – Britain had used individuals to gain information on hostile intentions of other countries, as of course had all major countries. Both the British War Office and the Admiralty had sections dealing with intelligence matters and both employed civilians as well as military staff. Some of the civilians were retired service officers. When they went abroad they were sometimes sent under journalist cover but in that case newspapers would have records of them : in the case of Meares there is no newspaper reference to him.

In 1895 the then Prime Minister, Lord Salisbury, had formed a Defence Committee to advise the Cabinet and Government on matters relating to the defence of the country and the Empire. This had no permanent secretariat or agents of any sort and relied on other organisations and departments for information. The debacle of the early days of the Boer War, particularly the complete absence of any proper evaluation of the Boer attitude to and competence in military matters, showed the need for something more efficient and focussed. In December 1902 a new Defence Committee was formed which in 1904 was re-named as the Committee of Imperial Defence and given its own permanent staff.

COMMITTEE OF IMPERIAL DEFENCE

The CID would have received information from War Office and Admiralty Intelligence and from individuals who reported on an ad hoc basis to Government departments of state and to British embassies and consulates in different countries. But there would have been no systematic or structured organisation within CID to oversee this.

In 1909 the British Government, and the Prime Minister Herbert Asquith who acted as the CID chairman, were concerned at the apparent placing in Britain of German agents. The CID stated there was no means of registering or investigating such activity and as a consequence – following a working group report – the CID recommended that a Secret Service Bureau be set up.

An Army infantry officer, Captain Vernon Kell, and a naval officer, Captain Mansfield Cumming, were then appointed to set up organisations within the Bureau to look at security, defence and intelligence issues within Britain and abroad. Later, in 1916, the two sections of the Secret Service Bureau became part of the newly appointed Directorate of Military Intelligence and were re-titled MI5 and MI6.

Meares's activities in the Russo-Japanese War could therefore not have been as a paid agent of the Secret Service Bureau or its sections because they were not formed until four years after the end of the Russo-Japanese War. What could have been the case was that Meares may have been approached by the British Embassy in China, Russia or Japan – or by one of the British consulates in any of the countries – to report on aspects of the Russo-Japanese War. Alternatively if he was not approached directly it may have been that his name was passed on by someone he met while travelling who then informed a British representative that he could be useful as a source of information. His ability to speak Russian – which developed while in Kamchatka – would also have been noted.

Meares was certainly involved to some degree in the Russo-Japanese War but the precise degree of that involvement remains unclear.

CHINA

After his six months travelling with Ponting he then journeyed to Tibet and China. It came about because of a chance meeting with a former Army lieutenant, John Brooke.

Brooke had served in the Boer War and – after leaving the Army in 1902 – had travelled in East Africa and then in March 1906 had sailed for India. He wanted to organise an expedition into Tibet to investigate the relation of the Sampo and Brahmaputra rivers but was prevented by the decision of the Indian authorities not to let any foreigners enter Tibet from India (as a result of pressure from the Russian government). He tried to slip across the Assam border into Tibet but was turned back. He then resolved to enter Tibet from the north via China.

Brooke went from Shanghai in June 1906 and reached Siningfu near the Tibetan border after a journey of three months. There he stayed for a month collecting supplies, ponies and yaks. While there he and the British head of the China Inland Mission, who was stationed there, met the Dalai Lama who had fled from Lhasa, the capital of Tibet after Sir Francis Younghusband's mission had entered Lhasa.

Brooke's journey into Tibet was difficult and he met with considerable obstacles from the local peoples and their leaders. Although he did establish good relations with a number, including a Mongolian tribal chief, and spent some time hunting and shooting Tibetan wildlife he never reached the Brammaputra river.

At the beginning of July Brooke returned to the house of two British missionaries in Siningfu and from there went back to Shanghai, this time via the Yangtse river. Then, later, in Tientsin he met Meares.

Meares described his meeting with Brooke in a speech he gave to an archaeological society in London [BC Ms0455] :

"In December 1907 I had just returned to Tientsin from a trip in Manchuria and was lunching in a hotel with the military attache when a tall bearded man came into the room and the attache told me that is Lieutenant Brooke who has just returned from Lhasa [the capital of Tibet]. I at once said how I would love to go there. As we were finishing our lunch Brooke came over to our table and asked if my name was Meares, I said it was, he said that he had heard of me and asked me to come up to his room.

"I went up and he opened a map of Asia and said we will start from here to Chentu, then we will explore the western tribes and thence to the Lolo country and west to the Brahmaputra. When will you be ready to start? I said next day."

The next day was Christmas Day 1907. Meares started by rail from Tientsin to Hankow by the banks of the Yangtse river and from there to Ichang, the highest point on the Yangtse, by steamer. There he hired a junk to take him to Chungking, the nearest point on the Yangtse to the small town of Chentu. Brooke joined him early in the journey and the two men, with their supporting party of local natives proceeded. Soon after leaving Ichang they reached the Yangtse Gorges where Meares described later "the river is hemmed in on both sides by cliffs more than a thousand feet high, and up on the face of the cliff a little ledge has been chiselled out of the rock along which the trackers crawl towing the junk". The Three Gorges is where the Chinese Government has now built the biggest dam in the world in order to bring and control water to the surrounding towns and villages.

Their junk was small and only had a crew of twenty people. When the Yangtse was calm, and the winds favourable, junks would generally go by sail but usually they were pulled by the crew. This was dangerous work. Meares related that junks "are towed by the crews who are harnessed to the end of a rope of plaited bamboo one or two hundred yards long, these ropes are wonderfully light and strong and they have to be as the weight of a heavily loaded junk in a strong current is tremendous and if the rope breaks the junk gets caught in the whirlpool and wrecked."

Nowadays tourists are taken for short distances up the Yangtse river in long canoes which are either paddled by river guides or else towed by ropes pulled by local people along the winding paths by the riverside. Although such travel today looks difficult it is nothing compared to the journey that Meares and Brooke made. It took them twenty-one days to go as far as they could up the Yangtse and had many very difficult and hair raising times with the river.

Meares stated in his lecture : "At places in the river there are very bad rapids where hundreds of extra coolies are hired and holding to these rocks with their hands and feet they work the junk up the rapids inch by inch, sometimes a rope breaks and the junk is carried off by the current and dashed to pieces on the rocks below. Once our junk was carried away and it looked as if we would be wrecked. The cook who was cooking a boiler of rice in the bow of the boat seized his ladle and heaved all the rice into the river. I asked him why he was doing it, he explained that it was to propitiate the river god. The junk drifted into a calm backwater and was saved, so the laugh was on his side."

He went on : "We tied up to the shore every night and saw much of interest, at one place there was a valley where they found numbers of bones of prehistoric animals. I saw the teeth of mammoths, sabre toothed tiger and others. The Chinese grind them up for medicine."

When they got to Chungking at the end of their river journey they hired some of the locals to carry their baggage and set off on the three hundred and fifty mile journey by land to Chentu. Meares described the beautiful country they travelled through, the farms, the colourful vegetation, fruit and flowers – particularly Himalayan poppies – they saw and the people they met. When they reached Chentu ten days after leaving Chungking he remarked later in his lecture to the archaeological society that "We thought that an average of over thirty miles a day was good walking, but it was rather humiliating to see our wretched coolies with 100 lbs of load keeping up with us quite easily."

In Chentu they met up with WN Ferguson. Ferguson and his wife had lived and travelled in the region of western China and near the Tibetan border for some years, mainly distributing books for the British and Foreign Bible Society, and knew many of the people who lived there. Ferguson – in his book 'Adventure, Sport and Travel on the Tibetan steppes' which was published in 1911 – wrote that although he had travelled extensively between 1903 and 1907 "very much still remained to be learned and Mr Brooke suggested that if he should decide to go into that province I should introduce him to those chiefs with whom I had already become friends". Ferguson agreed.

One of the reasons for Brooke and Meares travelling in the region was to hunt for animals peculiar to it – particularly the takin (a sort of Tibetan musk ox, also called a budorchus) and the goral (a type of Indian antelope). Meares was later to write about the different tribes of people they met and the animals they encountered and sometimes shot.

At one point Brooke and Meares did manage to capture a young takin and a goral. They thought they would take them back to a zoo in England but they died a few weeks after capture.

In the Badminton Magazine of October 1909 Meares wrote an article 'Reminiscences of a pleasant visit to a little known people' and in the December 1908 edition an article on 'Huntiing the Takin on the borders of Tibet'. In The Wide World Magazine editions of April, May and June 1910 he had three articles titled 'Among the unknown tribes'. He also wrote a booklet 'The Land of the Budorchus' which described the whole of the journey with Brooke and Ferguson and the killing of Brooke by men from the Lolo tribe. These were all deposited in the British Columbia Archives after the death of his wife in 1974 [BC Ms0455].

They all show considerable fascination with the country through which they travelled and the different tribes of people they met. The conditions they encountered were often very difficult with pouring rain, large areas of mud, sleeping in makeshift camps, climbing over mountain passes – some as high as sixteen thousand feet high – and often going where no European had ever been before.

Meares and Brooke with a Wassu Chief © Caedmon

At various times the three men – Brooke, Meares and Ferguson – split up and went their separate ways for a few weeks and then met up again. At one point Brooke and Meares had decided to meet again at Chentu and from there go south to what was described as 'Lolo country' and then enter India by the river Bramaputra. On their journey they came upon several cliffs and hills with caves – used both for habitation and for placing coffins in – which Brooke wrote about in an article he wrote shortly before his death 'Description of caves of Western Szechwan'; it was published the following year.

Both Meares and Brooke were fascinated by the caves. Indeed they are still very much a feature of tourist travels in China today. Meares also described their findings in some detail [lecture/archaeological society]:

"There are hundreds of prehistoric cave dwellings of great interest which are cut out in the red sandstone of which the hills are formed. These caves are of all sizes." He then went on to describe one of the larger ones.

"Some distance up the face of a cliff of red sandstone on the face of which carved hieroglyphics can still be seen, are approachable only by little footholds cut in the rock. A large cavern has been hewn out of the rock evidently by means of a metal instrument as the tool marks are still sharp and clear. In the cliff above the cavern a gutter is generally cut to shoot off the rain water and over the entrance imaginary animals are sometimes carved, perhaps the crest of the owner.

"This verandah would be thirty feet long by twelve feet wide by ten feet high and the roof is supported by two large pillars which were left when the rock was cut away. There was often an altar between the pillars. This verandah has square corners with a finely carved cornice running round the top. In the left wall of the verandah a small cave is often cut and in one I saw the stone figure of a dog. Generally three caves open from the main verandah, the one on the right about thirty feet long and quite straight was perhaps used for the servants or animals, the other caves were larger. All the caves had inner and outer doors, the first at the entrance, the second about twelve feet further in. There were grooves cut in the rock for the doors and socket holes to take cross bars.

"Half way along the larger caves are recesses cut in the wall containing large troughs cut in the solid rock and evidently used for holding water. A gutter was cut round the trough to carry off surplus water and the outer edge was much worn as if by constant use, in some of the troughs were round stones which showed signs of having been constantly heated in the fire. Just past the troughs were holes cut in the cornice evidently to take a curtain pole, and immediately past this the cave had been opened out into recesses large enough for rooms which also had sockets for curtains."

Meares and Brooke were both staggered by the caves and their contents. Many of them also contained coffins and must have been used as tombs. Several were also silted up with earth and, clearing this away, they found many terra cotta figures of extraordinary detail and beauty. Meares recalled that the local Chinese they met "said there was a curse on these caves and that misfortune would follow on those who entered them. Unfortunately this warning partly came true as most of those who entered the caves were killed soon after." He gave no further detail of this.

THE LOLO COUNTRY

They continued south for several days and reached Ningyuanfu, a large town on the edge of the Lolo country. There they waited for Ferguson to join them as they all three wanted to go into the Lolo country together. They then got a telegram from Ferguson saying he could not join them till later. In the light of that Brooke decided he would make a short journey to the Lolo border to get some photographs and see if the Lolo tribesmen were friendly. He told Meares that he hoped to be back in a few days – two weeks at the most. Meares was not to see him alive again.

Meares described the Lolos as "a race who have no resemblance to the Chinese and although their country is quite surrounded by Chinese territory yet they have never been subdued by the Chinese, as they are very warlike and live in very mountainous country. The Lolo women are tall and graceful and have olive complexions, they wear tam o' shanter caps, a silver collar round their necks, embroidered coats and accordion pleated skirts. The men have their hair twisted into a horn which sticks out from their forehead, they wear long felt capes, very loose breeches and sandals. Their weapons consist of bows and arrows with which they are experts, slings, beautifully made swords and very long lances. The chiefs wear armour beautifully made of lacquered scales of leather laced together and iron helmets inlaid with leather. They have good ponies and are expert horsemen, they have their own written language and many books but these are only understood by the priests. They are great drunkards and very bloodthirsty and spend their time fighting among themselves or raiding the Chinese in the surrounding country."

In fact there had been several bloody encounters between Chinese soldiers and the Lolos in the weeks just before Brooke set off.

Meares stayed behind to wait for Ferguson to catch up with them. But, as the days went by, Meares began to get more and more worried. He tried to send some Lolo tribesman into the Lolo country to try and find out what had happened to Brooke and his party. Several of the tribesmen came back with different rumours but the strongest was that Brooke had been killed. Eventually Meares found out precisely what had happened and related it to his audience.

"When Brooke reached the Lolo frontier he met a Lolo chief who was very friendly and made Brooke his blood brother. He invited Brooke to travel in the country and sent his son to escort him to the next chief, and he had a friendly reception wherever he went. In this way he travelled right across the country and could have returned through Chinese territory but as he was in a hurry to rejoin us he intended to return by a short cut. He got on very well till he reached the chief of the Aheo clan where [he] slept the night. During the night a number of things were stolen and in the morning Brooke refused to give them a present unless the things were returned. Brooke started off in the morning without recovering the things and all went well till at noon they were overtaken by thirty armed men of the Aheo clan who passed by and then waited for them in a narrow place.

"Brooke tried to parley with the chief who drew his sword and slashed Brooke's shoulder. He drew his revolver and killed the chief and fired at the others who bolted. Brooke then told his people to drop their loads and they hurried on till they came to the house of a chief called Suga who said that he would protect them and called them into his house. They were then surrounded by hundreds of Lolos and Brooke barricaded himself in the house and fought till his cartridges were finished. They then rushed the house and killed him." It was Christmas Day 1908 – exactly twelve months after Meares had first set off from Tientsin for Chentu.

Ferguson then joined Meares at Ningyuenfu. The local British consul asked Ferguson to go to Mapien where Brooke's body had been taken, to identify it and also bring it back to Chentu. The consul did not want the body of an Englishman left with Lolo tribesmen. Brooke's body was taken by some of the Lolo tribesmen to Mapien where they only handed it over after money had been paid to them.

Altogether, from when Brooke left Meares to go into the Lolo country to when Ferguson got to Mapien to retrieve Brooke's body was some eighty days. There he identified the body, saw it placed in a Chinese coffin, and with the help of some Chinese soldiers escorted the body back to Chentu.

Meanwhile Meares, having waited for some time at Ningyuenfu decided he would make his own way back to Chentu. His journey was uneventful except for when he met the Nepalese Ambassador at a small town on the way. He described his encounter in 'The Land of the Budorchus' [pp52-53] : "I arrived here in the evening and found that all the inns were full. I met some Indians in the street and asked them what was happening. They told me that the Nepal Ambassador was returning from Pekin [sic] to Nepal with 500 coolies carrying presents to Nepal, 150 picked ponies, 60 Ghurkhas, officers and men, as well as many Chinese officials and soldiers.

"I could get no quarter in the town so went to a small hovel higher up the street, and after my meal I went and called on the Ambassador, who I found to be a very pleasant old gentleman who spoke English perfectly.

"Next morning I started off my coolies before daylight, and on coming to the door to see them start, I found awaiting at the door, the Ambassador in a long silken kaftan and wound round his head was a sacred scarf which barely covered a golden plate bound to the top of his head. He was attended only by one Gurkha, who held a saddled horse.

" As soon as he saw me he threw his arms round me and began crying bitterly, saying he was dying, that the Chinese had conspired to poison him, that they had been putting poison in his food for several days, and that finally during the night they had tried to kill him with chloroform. He said that he threw himself on my protection and that I must take him to the British consul at Chentu. I induced him to come into my inn and tried to soothe him. After a few minutes some of his Gurkha officers came to look for him. As soon as he saw them he seized a pistol, which I had in my belt, and tried to shoot them. I took the pistol away from him and got him quieted again. After a time the Chinese official came to see him and he at once seized a large hammer which was lying near and tried to kill him. I just managed to catch him in time and got the official safely out of the house and tried to calm the Ambassador.

"When he got quiet I had a long talk with him and he told me many interesting things about himself and his country. I soon managed to slip out and told my servant to look after him. I then went to the telegraph office and sent a telegram to the consul at Chentu saying that the Ambassador insisted on returning with me to Chentu and asking him what steps I should take.

"After a time I got an answer back saying that the consul had left Chentu.

"I then went down to see his Gurkha officers and found these in a great state. They could only talk Hindustani, but fortunately I was able to talk to them and very nice fellows they were. They said that they were quite sure that the Ambassador had gone off his head.

"After a lot of talking we determined to telegraph to the Maharajah of Nepal asking for his instructions; so I wrote out a long telegram saying that the Ambassador had become strange and insisted on returning to Chentu and abandoning the expedition and asking for instructions.

"I took the telegram to the office and asked the telegraph clerk to send it, but he said he did not know where Nepal was or how much time the telegram would cost and that he would have to wait till he got particulars from Chentu.

"I then went back to the officers and got them to come up and see the Ambassador.

"They told him that no one had ever thought of trying to poison him and they all swore that they would protect his life. They also said that if he insisted on leaving the mission that they would never dare to return to Nepal and that they would all disperse over the world and become religious mendicants. After a lot of talk the Ambassador agreed to do what I recommended and finally agreed to continue his journey to Tachenlu where he would see how things got on. He also agreed to return to his house if I would come with him. So I took his arm and we marched down the town to his house and I stayed and talked with them till late at night. He gave me some photos of himself and some presents, and asked me to take any of the ponies I fancied, but I told him that I was leaving the country and did not need any ponies. They were most anxious that I should accompany them on their journey along the southern road to Lhasa, where they would stay for a month or more and then on to Katmandhu. It was the chance of a lifetime, but I had other work to do and had to refuse their offer and slipped back to my inn."

The next morning Meares carried on to Chentu and there stayed at Ferguson's house. Ferguson arrived from Mapien a few days later and then Brooke's body shortly after that. Brooke's body was buried in the foreign cemetery near Chentu, the service being conducted by an English priest from the Church Missionary Society.

TERRA NOVA

TERRA NOVA

Meares never wrote about his time with Scott's *Terra Nova* Expedition . In contrast to what happened after his travels in China and the borders of Tibet he wrote no articles for magazines or newspapers, nor did he make any speeches or give any lectures about his Antarctic experiences.

Several of the men who were on Scott's last expedition wrote books about it; many kept diaries. Meares kept no diary and his only comments about the expedition are to be found in a few letters to his father (written before he actually got to Antarctica) and what others recorded him as saying.

He had long wanted to go on a polar journey. When, after the end of the Russo-Japanese War, he was travelling with Ponting he remarked how exciting he thought a polar trip would be. This caused Ponting himself to consider such a journey. Then, in 1907, Ponting was travelling in Russia and he had with him a copy of Scott's account of his first expedition – 'The Voyage of the *Discovery*'. It convinced him that, if an opportunity arose, he should go to Antarctica as well.

Ernest Shackleton returned from Antarctica to London in June 1909. He received a hero's welcome. His *Nimrod* expedition of 1907 to 1909 had, after all, achieved three things – the furthest south of ninety-seven geographical miles from the South Pole, the attainment of the South Magnetic Pole and the ascent of Mount Erebus. He was received at Buckingham Palace, feted throughout London and awarded a knighthood by the King. On the crucial point, though, of actually reaching the South Pole – ninety degrees south – he had failed. Captain Scott knew it was now his one chance to achieve what Shackleton had failed to achieve, and to do it before anyone else could.

Robert Falcon Scott announced plans for his second Antarctic expedition in *The Times* and *The Daily Telegraph* on the thirteenth of September 1909. Offices for what was called the ' British Antarctic Expedition' were established in Victoria Street in London. Altogether some eight thousand applications to join the expedition were received.

Scott persuaded Dr Edward Wilson to join as head of the expedition's scientific staff. Wilson was his friend and Scott knew he could rely upon him completely. It was Wilson who had gone with Scott and Shackleton on the then furthest south journey on the *Discovery* expedition of 1901 to 1904. They had reached 82° 17' south but could go no further. The travelling surface was difficult, they could not handle their dogs, they were running short on food and Shackleton and Wilson were both ill – Shackleton so much so that Scott sent him home on the relief ship *Morning* in the summer of 1903.

Some of the other members of the expedition were recommended to him by Sir Clements Markham, the former President of the Royal Geographical Society, some by the Admiralty and others from those who were first recruited. Apsley Cherry-Garrard was one of those : he was recommended to Scott by Wilson and was originally accepted without even meeting Scott – though he had agreed to give one thousand pounds towards the expedition's funds.

A Royal Navy lieutenant, Edward (Teddy) Evans had been the second-in-command of the relief ship *Morning* and later had been thinking of organising his own Antarctic expedition. Clements Markham persuaded him instead to join Scott's expedition. He agreed and Scott made him second-in-command of the *Terra Nova* expedition.

As with most polar explorers at the time, one of Scott's initial – and indeed long lasting – problems was the shortage of money and the constant need to raise money for the expedition by appeals to wealthy individuals and the public. The Government had agreed to advance £20,000 towards the cost of the expedition and the Admiralty agreed to release those men from the Royal Navy who wished to go; but much more money was needed. Scott had estimated that the total cost of the expedition would be some £40,000.

Another who gave one thousand pounds towards the costs of the expedition was Lawrence Edward Grace Oates – Captain Oates, an Old Etonian and an officer in the Iniskilling Dragoons. The Iniskillings, a heavy cavalry regiment, had been sent to South Africa as part of the British Army forces in the Boer War and landed in South Africa in January 1901. Oates was a junior officer with them. He had considerable experience of riding and had particularly wanted to join a cavalry regiment.

He was involved in an encounter with Boer forces by a small hill near the town of Aberdeen in Cape Colony, when his small detachment was surrounded; he was shot in the thigh but refused to surrender. He and his men successfully resisted until the Boers retreated and for his action he was mentioned in despatches and, it was reported, was also recommended for a Victoria Cross. The bullet that shattered a bone in his thigh left his left leg slightly shorter than his right and this was to cause him difficulty on the return journey from the South Pole eleven years later.

Oates returned to England to recuperate and then went back to South Africa and remained with the Iniskillings until the end of the war. Thereafter he served with his regiment in Egypt and India. From India he wrote applying to join Scott's expedition and, after supplying the requested references, was accepted. He became a close friend of Meares on the expedition.

Meares had applied to join the expedition and an official at the Admiralty had also recommended him to Scott. He was accepted as a transport officer and put in charge of the dogs. Meares then in turn recommended to Scott that Herbert Ponting should join as the photographer – or, as later described, 'the camera artist' – and Scott, having heard of Ponting's work and reputation, agreed. Ponting had been planning a two year tour of different countries of the British Empire under contract to the Northcliffe Press but his interview with Scott in the autumn of 1909 had convinced him to join Scott's expedition.

Scott decided that there would be four methods of travel for the expedition in Antarctica – sledges pulled by dogs, by ponies, by motor tractors or by men. He arranged for three motor tractors to be taken and recruited Bernard Day, the motor expert with Shackleton's *Nimrod* expedition, to be in charge of them. Meares would select the dogs and bring them from Siberia straight to New Zealand : Oates would select the ponies and would also bring them to join the expedition.

Meares went in January 1910 by the Trans-Siberian railway to Khabarovsk in Siberia. From there he went by horse and sledge some six hundred and sixty miles by the frozen river Amur to the small town of Nikolievsk near the Sea of Okhotsk. He was helped there in his choice of dogs by the English manager, Rogers, of the local branch of the Russo-Chinese Bank. From there he wrote to his father in a letter dated the eighteenth of March 1910 [SPRI] :

"Many thanks for your post card, received some time ago. My permanent address which will always find me is c/o British Consul, Vladivostok.

"I have been kept very busy collecting dogs, trying teams and picking out one or two dogs and making up a team and trying it on a run of 100 miles and throwing out the dogs which do not come up to the mark and collecting others.

"I have been having glorious weather here, bright sunshine and hard frost with a blizzard now and again for variety, There is a great degree of snow here, it lies 8 – 10 feet deep in the streets. I have been for a number of interesting trips and seen some interesting things. I have taken a number of photos and will send you some as soon as I can get some papers.

"I would be very glad to get some letters or papers, not post cards, as I have had no news since I left England and have no idea what is happening.

"The road here will soon be broken so I will not be able to send or receive any more letters till the spring. I expect to be back in Vladivostok by the middle of June where I will collect the ponies and I hope to leave Kobe for Australia by the beginning of August.

"It is a very big contract indeed to choose all these animals and carry them down to Australia single handed. I hope they will get down all right. I have so much to do that I must stop. With kindest regards to all. Your affect [sic] son Cecil H Meares."

Scott's decision on transport has been the subject of much argument for one hundred years. His own experience of dogs on the *Discovery* expedition had been disastrous. Neither he, Wilson nor Shackleton, had any experience of handling them. The dogs continually fought among themselves and Scott very much disliked the necessity of killing some, either for food for the other dogs or because they were too weak to go any further – in fact all either died or were shot before they returned back to the *Discovery* after their furthest south journey. But because of advice – particularly from the Norwegian explorer Fridtjof Nansen whom he visited in Norway in the summer of 1909 – he decided he would take thirty dogs (in fact the actual number of dogs originally shipped was thirty-one). It was not clear at first, however, what part they would play in the southern party : a matter which was to exasperate Meares. It was in contrast, too, to the Norwegian explorer Amundsen's decision to take one hundred dogs.

Scott was impressed by the fact that Shackleton had taken ponies on his southern journey and that he had got close to the South Pole. What, perhaps, he failed to appreciate was that it was taking ponies and not dogs that actually had lessened the chances of Shackleton reaching the Pole. As it was, he was also impressed by the fact that the ponies which had fared best with the Shackleton party were white ones and not black ones. He wanted twenty ponies in all.

One of the advocates of taking ponies was Albert Armitage, the navigator and second in command of Scott's *Discovery* expedition. He had previously been the second in command of the Jacksaon-Harmsworth expedition of 1894 to 1897 to Franz Josef Land in the Arctic Ocean, north of Russia. He thought [SofA EHp190] that, on a level surface like the Antarctic Ice Barrier, ponies would do better than dogs. A Siberian pony could stand severe cold and drag a heavier load in proportion to its weight and the amount of food it needed than dogs could do. Furthermore, pony meat would be more acceptable to eat than dog meat.

What Armitage omitted from his advice was that dogs could eat food found in Antarctica – seal and penguin meat – were not so heavy as ponies and therefore less likely to fall through thin snow and ice or down crevasses and, when food ran out, they could eat each other. But Scott's mind was made up : and if the motors, the ponies and the dogs were found to fail, they would rely upon man-hauling, something which Scott more or less looked forward to.

Oates, on arriving in England from India, was sent almost immediately to the West India docks in London to assist in helping with the overhaul and the loading of stores and equipment on to the *Terra Nova*, the old Dundee whaling ship that Scott had purchased for the expedition. Apparently Lieutenant Teddy Evans, Scott's second in command of the expedition, and Lieutenant Victor Campbell, who was to lead the Northern Party of the expedition, were so impressed with Oates' enthusiasm and effort on the *Terra Nova* that they both asked Scott to let him stay with them to complete the multitude of tasks necessary to get the ship ready. This meant that Oates would not be able to join Meares in Siberia and select the ponies. Scott agreed.

Meares knew about dogs. On one of his earlier visits to Siberia he had travelled across Siberia to Cape Chelyuskin, one of the most northerly points on the Russian mainland, by dog sledge : a journey of some two thousand miles. However, while he also knew something about cavalry horses from his time with the Scottish Horse, he knew nothing about Siberian ponies. In the absence now of Oates he had no choice other than to choose them himself.

Before leaving Nikolievsk with the dogs he had persuaded a young dog driver, Dmitri Gerov, to accompany him to Vladivostok and then join Scott's expedition. At Harbin, where he bought the ponies, he persuaded a Russian jockey from the Vladivostok race course, Anton Omelchenko, also to join him.

One of the problems at Harbin was that there were not too many white ponies for sale. Eventually Meares – rather than choose the ponies himself – asked an Australian horse trainer there to choose them with him and, settling on twenty, he paid over the money to the dealer. He paid the equivalent of five English pounds for each pony. Omelchenko was later to say that the horse dealer departed "with a plenty big smile" [Deb 18 June 1911].

To get the ponies and the dogs to New Zealand was a difficult task. Meares cabled Scott from Nikolievsk asking if someone could be sent out to him to help in the transport. Scott asked Wilfred Bruce, the brother of his wife

Kathleen, a Royal Navy Reserve lieutenant and chief officer on a P & O steamer, to go. Bruce had been travelling in the Far East and got back to England in thirteen days, via the Trans-Siberian railway. He described what happened much later [TBP volXII June 1932] :

"I arrived in London on June 13 and went at once to see Scott ... he went on to explain to me that he had tried to stop me on the way home, as Cecil Meares, who had been sent to Siberia to collect ponies and dogs for the expedition, had asked for another man to transport them from Vladivostok to New Zealand.

"Captain Lawrence Oates was eventually to take charge of the ponies, and Scott had intended to send him out to Meares. But Oates very much wanted to sail all the way out in the *Terra Nova*, so Scott asked me if I would mind taking his place as the long sea voyage would probably be no attraction to me. I suggested that if I could have two or three weeks in England, the return journey by rail across the continent would be of no great hardship and the plan was settled at once."

Bruce – after his leave in England – then went back via the Trans-Siberian railway again and joined Meares in Vladivostok, leaving England on the ninth of July and arriving there on the twenty second of July. He wrote "Meares met me in the train ... and took me to a small hotel which has nothing to recommend it except quite a luxurious bathroom, which after thirteen days in a hot and dusty train, was very enjoyable.

"We had all our meals together in restaurants. I found he was fairly well known there, and he introduced me to many of his friends, with some of whom, however, I had no language in common.

"He took me at once to see the twenty ponies and thirty one dogs he had collected up country, and with whom he was quite pleased. He had managed to get the dogs down on a Russian naval destroyer, by judicious handling of the commander.

"The dogs at that time were exceedingly fierce and wild, which rather surprised me, as Meares told me they had nearly all been driven in mail sledges in the north, and were well trained to harness. Before I arrived two of them had managed to get loose, and had dragged down and killed a horse before they were secured."

They then put the animals on to a small Japanese steamer, *Tategami Maru*, an experience which Bruce later described as "dreadful. Rain was falling in torrents, the streets and quays many inches in mud ... We had started the shipment at about 7 a.m., thinking we should finish in about two hours, and then have breakfast. It was after 4 p.m. when we got our first meal, wet through to the skin, and absolutely covered in mud from head to foot."

Bruce then told of their journey to New Zealand "The ship left next day, on a very leisurely voyage, calling at four ports in Corea [sic], and arrived at Kobe [in Japan] on August 4. By then Meares had emptied his purse, but I was well known here, and had no difficulty in obtaining the necessary money to carry us on.

"Here we had to tranship the ponies and dogs to another vessel. No British shipping company would take us, so we left two days later in the German steamer *Prinz Waldemar*. She was a passenger ship, and we were far from popular on board. I must own, though we did our best to keep everything as clean as possible, the dogs were far from savoury, and quite frequently would howl in unison in the middle of the night, keeping it up for quite a long time. Our unpopularity can, therefore, be understandable.

"The ship was very slow, the voyage not very interesting, but the weather was fine, which was a great asset, and our fellow passengers cheerful company.

"We called at Hong Kong, Manila, a little coral island called Yap, where there was a German wireless station, several ports in New Guinea, Raboul, Rockhampton and Brisbane and reached Sydney on September 9.

"We transhipped all our ponies and dogs again to the New Zealand steamer *Moana* ... we left next day for Wellington, and again were lucky in our weather and arrived there on the fourteenth. Once more we had to change ships, and joined the *Maori*, for Lyttleton, sailing the same evening."

THE PONIES

He described the difficulties of transporting the ponies : "We had become experts at the business by this time, but the ponies appeared to get more and more frightened on each occasion. We had to blindfold them now before they were hoisted out of or into a ship, and as I was covering up the head of one in Wellington, he struggled so much, and threw his head about so quickly, that I arrived in Lyttleton next day with black eyes and a swollen nose.

"The ponies had now been on their legs for fifty two days, as we never allowed them to lie down. It seemed cruel, but all the experts were adamant that it was the right thing to do. Sometimes a pony seemed to be distressed by the slight movement of the ship, we passed a band under him, but it was never very successful."

They landed the ponies and the dogs at the quarantine station of Quail Island, five miles off Lyttleton. The next month – on the twenty ninth of October – the *Terra Nova* reached Lyttleton. It had taken Meares nine months to select the dogs and ponies and transport them down to New Zealand when the ship arrived.

Meares, though, was to relate their journey – and cast the usefulness of Bruce – in a different light. In a letter to his father from the Caroline Islands [22 August 1910, SPRI] he wrote : "Just a few lines to let you know that I have arrived so far safely with my menagerie and they are flourishing up to the present. I had a job shipping them at Vladivostok. It was pouring intently with rain and the mud was two feet deep. However I managed to get them all on board safely but the forage got wet and damaged. At Kobe I had to tranship the whole bunch and the German steamer was 24 hours late so I had to keep them in open lighters in a blazing sun which was very hard on them.

"Capt. Scott's brother-in-law came out to help me, he is chief officer on a P & O steamer. Quite 'one of the boys' but too 'kid glovey' for this job, he stands on the upper deck and looks on instead of taking off his coat when there is a hard job of work, so it keeps me pretty busy. I generally go round once or twice in the night and then begin work at 5.30 a.m. ... We are fortunate in having fine weather which help to keep the animals fit ... The expedition will last longer than we thought at first probably 3 ½ years."

Meares' journey in getting the animals to New Zealand was remarkable. Only one pony and one dog were lost on the way down. It was a punishing time for both Meares and the dogs and ponies and that was before the expedition really started. Scott was not to know that Meares' journey had been successful until he went to Quail Island after arriving at Lyttleton.

Meares was never friendly with Scott and his dislike of Bruce was to endure to the extent that, after they had reached New Zealand, they remained on non-speaking terms. Bruce, as one of the ship's officers, sailed to Antarctica with the *Terra Nova* and returned with the ship to New Zealand after the shore party had been landed.

The *Terra Nova* stayed at Lyttleton for just under one month. The whole cargo had to be taken out and re-stowed, the ponies and dogs put on board, stalls for the ponies built, coal, food and equipment stored. Finally the ship sailed to Port Chalmers, by the southernmost New Zealand city of Dunedin, and from there to the Antarctic on the twenty ninth of November 1910.

Scott had travelled separately from England to Cape Town in South Africa on the mail steamer *Saxon*, together with his wife and the wives of Dr Wilson and Lieutenant Evans. There he joined the *Terra Nova*, took over as captain, much to the chagrin of Evans, but left the ship after it docked in Melbourne. He then proceeded separately by a passenger steamer to Lyttleton and rejoined the *Terra Nova* there. It was when in Melbourne that he received a telegram which Roald Amundsen had sent from Madeira : 'Beg leave to inform you. Am proceeding Antarctic. Amundsen'. It meant that there was now a rival expedition to be the first to reach the South Pole.

Scott had always maintained that while he hoped to be the first at the South Pole, his expedition would also carry out a lengthy programme of scientific work and also send parties to different parts of the area where the *Terra Nova* would go.

Amundsen, on the other hand, having switched from his original plan of putting his ship, the *Fram*, into the ice of the Arctic Ocean by the top of the Bering Strait, and drifting to the North Pole (an adaptation of Nansen's earlier voyage in the *Fram* from 1893 to 1896) made it clear that the single aim of his expedition was to reach the South Pole. When at Madeira he announced this to the men of his expedition, who had joined to travel north, his men agreed to go. Amundsen's brother, Leon, then announced this to the press when he returned from Madeira to Christiana [Oslo].

NORTH POLE

The Americans, Robert Peary and Frederick Cook, had both announced they had separately reached the North Pole – Peary in April 1909 and Cook in 1908 – and there would be little point in Amundsen now trying to go there.

In fact the current consensus among polar explorers and historians is that neither Peary nor Cook actually got to the North Pole and the controversy over that has lasted for a hundred years and still continues. It now seems likely that the first surface crossing of the Arctic Ocean to reach the North Pole was by Wally Herbert's expedition of 1968 to 1969. If Amundsen had kept to his original plan then , if successful, he would have been the first man there. In fact, Amundsen probably was one of the first men to see the North Pole. In May 1926 he – and fifteen other men – flew in an airship from Spitzbergen to Alaska : it was the first flight over the Arctic Ocean via the North Pole.

When Scott arrived in Lyttleton he went straight to Quail Island to see the ponies and dogs and pronounced himself satisfied. Oates, who had also travelled down on the *Terra Nova*, however, was appalled at the condition of the ponies. In his diary he wrote of some of the individual ponies "narrow chest, knock knees, aged, windsucker, doubtful back tendons off fore legs, slightly lame, pigeon toed, ringworm, suffering from debility and worn out" and concluded "I have only mentioned those which appear to actively interfere with their work or for identification."

Oates, though, did not blame Meares for their condition. He knew that he should have chosen the ponies and that Meares was not properly qualified to do so. Oates' concern was with Scott. Apart from the condition of the ponies he argued with Scott over the amount of fodder the *Terra Nova* could take; eventually Scott had to give way a little and increased the amount – but at the expense of coal.

The *Terra Nova* was built in 1884 and at that time was the biggest whaling ship afloat. It was a wooden three masted square rigged barque of seven hundred and forty-seven tons with an auxiliary steam engine and a reinforced iron bow. It had been one of the two relief ships sent to the Antarctic by the Admiralty, to assist Scott's *Discovery* in 1904. Scott had wanted the *Discovery* again as the expedition ship but the chairman of its owners, the Hudson Bay Company, – Lord Strathcona – had refused.

Scott had paid five thousand pounds for the *Terra Nova* with the promise to pay a further seven and a half thousand pounds at a later date. It had been in a poor and dirty condition when bought but was overhauled in London : again, however, time – and cost – had been a problem and not everything that should have been done was done. The pumps had not been replaced and this became a serious problem when the ship met a fierce storm just three days after leaving New Zealand for the Antarctic.

On board was Captain Scott with six Royal Navy officers (including Henry Bowers who was actually in the Royal India Marine), twelve scientists (including Meares and Ponting and also Trygve Gran, a Norwegian ski expert whom Scott had been persuaded to take by Nansen) and fourteen others – all Royal Navy seamen except for the two Russians, Gerov and Omelchenko. The ship's party was thirty two men with Commander Harry Pennell of the Royal Navy now taking over as the ship's captain.

There were therefore sixty-five people on board, nineteen ponies, thirty dogs and the pieces of the hut to be erected when they landed. The three motor sledges were housed on deck and there were stalls for the ponies The ship was overloaded. Fortunately, in a way, before leaving England Scott had been elected as a member of the Royal Yacht Squadron and thus was able to fly the RYS pennant : but it also meant that the *Terra Nova* was not subject to Board of Trade regulations on ship safety. Indeed Evans, captain of the ship when it left England, actually had the plimsoll line painted over [SwS p9].

Three days after the ship had left Port Chalmers a big storm blew up with a force ten gale. The pumps did not work properly and, until they managed to clear them, all the men had to bail out the water with buckets. Several times the ship nearly listed over but after thirty six hours the storm subsided. Two of the ponies died during the storm.

Seven days later, on the ninth of December, the *Terra Nova* met the first big ice floe and some icebergs : it was much further north than when the *Discovery* had sailed down nine years earlier. The next day they crossed the Antarctic Circle.

Then, on the thirtieth of December, the ship was free of the pack ice and into the open water of the Ross Sea by the Ice Barrier. They had been in the ice for twenty days compared to the four days the *Discovery* had spent in the same waters. It was one of several occasions that Scott was to describe as bad luck.

Three days later Scott could see the outline of Mount Erebus as they entered McMurdo Sound. Ice blocked the way to anchoring where the *Discovery* had been anchored before at Hut Point, so they landed about six miles north of Cape Royds (where Shackleton's *Nimrod* expedition had erected their hut). The place where they landed had been called the 'Skuary' but Scott renamed it 'Cape Evans' after the second- in- command of the expedition. It was the fourth of January 1911.

By the eighteenth of January the hut was erected and all the equipment, stores, dogs and ponies had been off loaded – Meares, Gerov and the dogs had been invaluable in helping to transport all the material from the ship.

THE DISCOVERY HUT

Three days earlier Scott and Meares went with a dog team to inspect the old *Discovery* hut at Hut Point. Shackleton's party had used it several times during the *Nimrod* expedition and Scott was shocked at the mess that he and Meares found – mainly because Shackleton's party, after their desperate journey back from their furthest south, had been in a great rush to get back on board the *Nimrod* and sail away. A window had been left open and there was a lot of snow inside.

Scott wrote in his diary that "Boxes full of excrement were found near the provisions and filth of a similar description was thick under the verander [sic] ... It is extraordinary to think that people could have lived in such a horrible manner and with such absence of regard for those to follow". Those remarks in his diary were subsequently deleted when his journal was published [AFRT p139].

A disaster, however, had occurred four days after they had begun unloading the ship. Two of the three motor sledges had been successfully unloaded but Scott – 'stupidly' as he put in his diary for that day – had agreed the third motor should be unloaded when the ice was melting around the ship. It sank and was lost.

Scott had brought the motors with him to help with the dogs and ponies but not to replace them. They were petrol driven with a four cylinder, air cooled, twelve horse power engine and made by Wolsey Motors of Birmingham and only capable of a maximum speed of three and a half miles an hour. Motor sledges had never been tried in the Antarctic before (Shackleton's *Nimrod* expedition had a motor car – but that also had proved unsatisfactory). It was a brave experiment to take them but now Scott was down to two motor sledges before any sledging had been undertaken.

Scott's task was now to lay depots before the Antarctic winter set in and again Meares and the dogs proved their use. At first, though, Scott expressed doubt about the dogs. In his diary [24 Jan 1911] he wrote "Wilson and I drive one team [of dogs] while E [Teddy] Evans and Meares drive the other. I withhold my opinion of the dogs, in much doubt as to whether they are going to be a real success – but the ponies are going to be real good".

Surprisingly, though, three days before they started on the main depot laying journey Scott wrote to Sir Joseph Kinsey, the expedition's agent in New Zealand. The *Terra Nova* would take the letter back to New Zealand.

In the letter [22 Jan 1911] Scott asked if Kinsey would arrange to "send down a team of Indian transport mules on the ship next year". He also wrote to Major-General Douglas Haig, who was then Army chief of staff in Simla in India, requesting the mules : "I have thoroughly discussed the situation with Captain Oates and he has suggested that mules would be much better than ponies for our work and that trained Indian transport would be ideal ... Oates and another member of my expedition, Mr Cecil Meares, have seen the wonderful work done by mules in northern India, and especially during the expedition to Thibet [sic]". For the moment, however, on the southern journey Scott had to use the ponies he had.

In fact the mules were sent with the *Terra Nova* when the ship returned to the Antarctic at the end of 1911 and were taken by the men who set out to find the bodies of Scott's final Pole party in November of 1912.

Scott's hut at Cape Evans was at 77° 38' south. It was thirteen miles north of Hut Point. The problem was that the thirteen miles were over the frozen ice : journey by land from Cape Evans to Hut Point was difficult, if not impossible, for the dogs and ponies. It was all right when the sea was frozen and they could travel over the ice but, when the ice melted, they hardly could get from one to the other.

ONE TON DEPOT

On the twenty fifth of January, just twenty one days after landing at Cape Evans, Scott led the main depot laying party of twelve men, eight ponies and twenty six dogs. He, Wilson, Teddy Evans and Meares shared one of the four man tents. His intention was to lay depots up to 80° south and there place a big load of supplies to be used by the southern party on their return from the South Pole. They called it One Ton Depot. Amundsen – landing at the Bay of Whales, some sixty miles further south than Cape Evans – in contrast, laid depots at 80°, 81° and 82° south and at one point had suggested laying a depot at 83° south but later dropped this idea.

The first depot was laid by Scott's party a few miles from Hut Point at a spot where Scott estimated, rightly, that the ice would not break away from the shore when spring and summer came. They called this 'Safety Camp' and then thirty miles south from Hut Point they laid another depot – Corner Camp. From there Scott estimated they would march for about ten days to reach 80° south.

While the depot laying party was going south, five of the shore party, including the two Australian geologists, Frank Debenham and Griffith Taylor, went to study three glaciers west of Ross Island on Victoria Land, a journey that was to take six weeks. The *Terra Nova* also left the base at Cape Evans to land Victor Campbell, Raymond Priestley, the other Royal Navy surgeon, Murray Levick, and three others – Frank Browning and Thomas Williamson, both Royal Navy petty officers and Harry Dickason, an RN able seaman – on King Edward VII Land. They were named the Eastern Party.

AMUNDSEN

Unable to find a safe place to land the Eastern Party on the Ice Barrier, Pennell decided to see if it was possible to land in the Bay of Whales. Shackleton had previously decided on the *Nimrod* expedition that it was unsafe to land there because of the breaking away of the ice into the sea and he was to incur Scott's wrath when he had subsequently established his base at Cape Royds in McMurdo Sound. Pennell, though, took the ship into the Bay of Whales and was astonished to see there that Amundsen's ship, the *Fram*, was anchored by the ice.

Pennell, with two others from the ship, went across to meet Amundsen and breakfasted on the *Fram*; then they were taken to Amundsen's hut, met the other Norwegians of the party, saw the dogs and in turn took the Norwegians back to the *Terra Nova* for lunch. During their rather guarded conversation

Amundsen told Pennell that the *Fram* had only taken four days to get through the pack ice to the open Ross Sea – compared to the three weeks that it had taken the *Terra Nova*.

Afterwards, Pennell decided he had to go back to Cape Evans to leave a letter for Scott on what he had learnt about Amundsen's base. Then, as the Eastern Party agreed they could hardly land at the Bay of Whales, they decided the *Terra Nova* should land them on the other side of McMurdo Sound. So the *Terra Nova* took them to Cape Adare, to the north west of Cape Evans, where the hut of Carsten Borchgrevinck's 1898 – 1900 British expedition was and where Borchgrevinck and ten men were the first men to winter on the Antarctic mainland. Campbell's party was now named the Northern Party. The intention was that the party would explore and survey the area. They had wood with them to build a small hut, and the ship would pick them up in the Antarctic summer, probably in January or February 1912.

In the meantime, at Corner Camp, Scott's depot laying party was held up by a blizzard for three days. After that they went forward for ten days. Then Scott decided that three of the ponies were so weak that they would be sent back with Lieutenant (Teddy) Evans and the two Royal Navy petty officers, Forde and Keohane. The others would go on – five men with the other five ponies and Meares and the other three men with the two dog teams. Cherry-Garrard later wrote [WJW p340] that the dog teams started after the men with the ponies and caught them up "since they travelled faster than the ponies".

On the fourteenth of February it was Meares' birthday. In his diary for that day Wilson wrote : "As this was Meares' birthday we had a special effort for the supper meal consisting of dry hoosh – i.e. biscuit crumbs fried with pemmican and cheese rind and curry powder – thickened with arrowroot and sultanas and a piece of chocolate".

By the seventeenth of February the party had gone as far as Scott considered they should go. They had gone no further than twelve miles a day; several days averaging just over six miles. It was taking longer than Scott had hoped. The conditions and the temperatures meant that the five remaining ponies were getting progressively weaker. He thought that if they went any further the ponies would succumb and have to be shot – and he needed them for the main southern journey.

Scott decided – against the wishes of Oates, who thought they should have pressed on as the ponies might die anyway on the journey – that they should now lay the One Ton Depot. They had reached latitude 79° 29' south – some thirty miles short of the eightieth degree. It was a fateful decision in view of what was to happen to Scott's party on their journey back from the South Pole.

Scott then decided he, Wilson Cherry-Garrard and Meares would go back with the dogs, leaving the others to follow with the ponies. Scott was anxious to see if Teddy Evans' party had got back all right and what the conditions of their ponies was and also to see what news there was of Victor Campbell's party.

They made good progress, going up to thirty miles in one day. When they were about twelve miles from Safety Camp Scott decided they could try a different way forward, but the area was heavily crevassed. Wilson described what happened in his diary [EW] :

"I was running my team [of dogs] abreast of Meares, but about 100 yards on his right, when I suddenly saw his whole team disappear, one dog after another, as they ran down a crevasse in the Barrier surface. Ten out of his 13 dogs disappeared as I watched. They looked exactly like running down a hole – only I saw no hole. They simply went into white surface and disappeared. I saw Scott, who was running alongside, quickly jump on the sledge, and I saw Meares jam the brake on".

They had been running on the thin surface of a crevasse about six to eight feet wide. One of the three dogs who had not fallen in seemed to stop the harness pulling the others down. The sledge was still on the surface. Wilson and Cherry-Garrard stopped their sledge and went over to assist. Meares was lowered down on an alpine rope and managed to tie the end to the harness. They pulled up the dogs but two, who had somehow slipped out of the harness, were about forty feet down on a ledge.

Wilson's diary recorded that "after testing the length of the alpine rope we let Scott down at his request. We all wanted to go instead of him, but he insisted and so we recovered all the dogs that had gone down."

Cherry-Garrard wrote [WJW] a similar story but, curiously in his later 'Annotated Journals' [FS p214] stated that "Scott told Meares to go down and get the dogs. Meares refused. I said I often went down the well at home let me go! Scott said to Bill [Wilson] 'What do you think?' Bill said he didn't think anyone should go, but if anyone went he would go down. Scott then said he [would go] down; and he went."

There is no other statement to back up what Cherry-Garrard wrote about Meares' alleged refusal. Scott in his diary made no mention of it nor did Wilson. Cherry-Garrard wrote several years later – again in the Annotated Journals – that "Up to this day Scott had been talking to Meares of how dogs would go to the Pole. After this, I never heard him say that."

Meares making harness for some of the dogs he brought from Siberia.

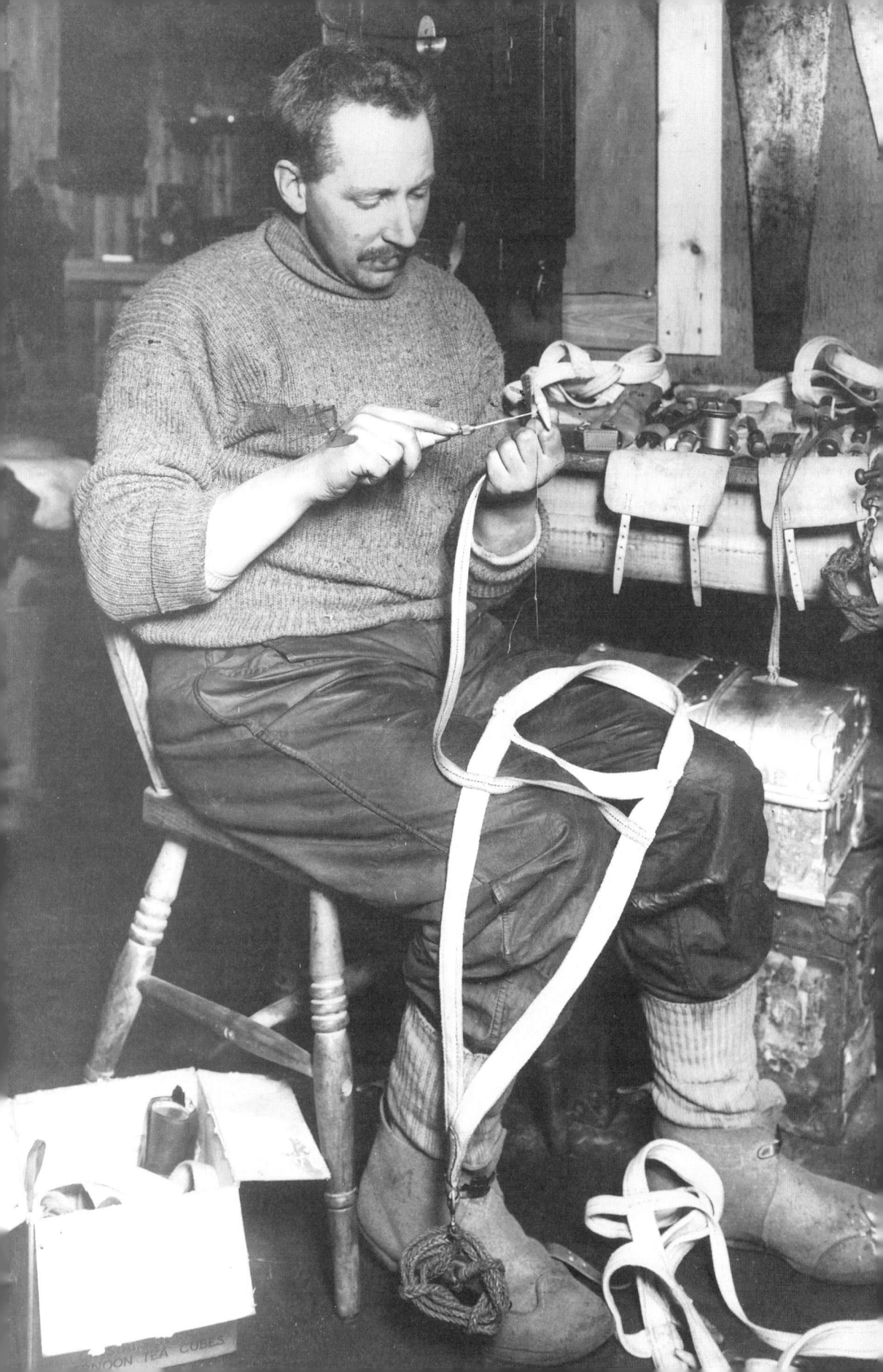

It would be very surprising if Cherry-Garrrard's statement is correct as there is no trace anywhere – in the diaries or books about Scott's expedition – that Scott ever expressed the view that the dogs might go to the Pole.

The next day they reached Safety Camp. There they found Evans, Forde and Keohane but with only one of the ponies : the other two had died on the way back. Then, because there was open water by Hut Point, Scott, Evans, Wilson, Cherry-Garrard and Meares decided to go there by the land route, on foot and with one sledge. They got to the hut but found no signs of news there, except for a note which stated 'Mail for Captain Scott is inside bag inside south door'. They could not see any bag, though. Scott then saw some foot tracks outside and realised that the surgeon, Edward Atkinson – his foot apparently having healed – and Tom Crean, one of the other Royal Navy petty officers, must have returned to Safety Camp. They had left the depot laying party almost at its beginning because of a damage to Atkinson's foot and Crean had assisted him back.

They went back to Safety Camp and there was the letter from Victor Campbell which Pennell had left at Hut Point when the *Terra Nova* had called there on its way to Cape Adare with the Northern Party. It had been taken to Safety Camp by Atkinson and Crean to ensure that Scott would receive it as soon as possible.

Campbell had written about the encounter with Amundsen and his impressions of the Norwegian's base. It was a bitter blow. Scott had previously surmised that Amundsen would go to the other side of Antarctica, by the Weddell Sea, and set up his base there. He now saw that Amundsen's base was sixty miles nearer the South Pole than his own and that the Norwegian had about a hundred dogs with him. And Amundsen could set out on to the Barrier from his base and not – as Scott would have to do – travel the more difficult journey from Cape Evans to Hut Point and then through to the Barrier.

The next day Wilson and Meares went with the dogs to get some seals to feed the dogs and to keep some for the party. The others left Safety Camp to go to Corner Camp and await Oates, Bowers and Gran with the three ponies they were bringing back. They met up with them and they all returned to Safety Camp.

KILLER WHALES

At the end of February Scott decided they should all try and get back to Hut Point. Meares and Wilson went off with the dog teams. They managed to get across by the land route but it was too difficult to get the ponies that way.

The others had to take the ponies by the sea ice route. Their journey was a nightmare, with the ice floes breaking up and then they encountered a number of killer whales. The ponies were weak and scared of the whales. It was impossible for the ponies to jump from one broken floe to another and several had to be killed rather than leave them to the mercy of the whales. Eventually they got to Hut Point but of the original eight ponies that started with the depot laying party in January now only two were left.

They were all at Hut Point by the fifth of March. They could not cross from there to Cape Evans until the sea ice became frozen and strong enough to bear their weight. Ten days later Debenham and his party that had gone to do geological survey in Victoria Land, returned to Hut Point.

There the whole party stayed for a further five weeks. It was a difficult time because the *Discovery* hut had never been intended for a lengthy stay and food was running short. In his previous expedition Scott and his men had stayed on board the *Discovery* and only used the hut for storing supplies and equipment.

Then on the eleventh of April Scott and half the party left to get to their base at Cape Evans. Meares and the remaining men of the party stayed until the ice was thick enough to take the dogs and ponies back. They were to wait for another month.

Having got to Cape Evans, Scott and a small party, on the eighteenth of April, went back to Hut Point, over the land route, to bring supplies for the men there. Then, on the twenty-first of April, Wilson, Cherry-Garrard, Atkinson and Oates went back to Cape Evans with Scott's party. This left Meares and the remaining four men at Hut Point.

By the second week of May the sea ice had frozen hard enough for Meares and the others to bring the dogs and ponies back to Cape Evans. And they made it without mishap. They had originally left Cape Evans with the depot laying party towards the end of January. Now, three and a half months, later they were back at the base hut.

On the twenty third of April the sun went below the horizon for the last time and the Antarctic winter set in. There was a lot to do – adjustments to the sledges, putting food into separate bags for the southern journey, measuring the wind and temperatures, and mending and adapting their equipment, boots and clothing.

Victor Campbell's Northern Party spent the winter in their small hut by Cape Adare. The *Terra Nova* picked them up early the following January and then landed them further along the coast in Evan's Coves. The coves had

been first sighted by the *Nimrod* when sailing along the coast to search for the South Magnetic Pole Party of Shackleton's 1907 - 09 expedition. They were on the edge of an island which Campbell's party called 'Inexpressible Island'. They were to stay there doing geological and meteorological work until the *Terra Nova* picked them up again in the middle of February. In fact the *Terra Nova* was prevented by the thick ice – and the shortage of coal – to get to the coast and pick them up although three attempts were made. Campbell's party, therefore, had to spend another winter away from the base at Cape Evans – this time in a dug out cave.

The story of the Northern Party, written later by the geologist Raymond Priestley, is itself a masterly account of great privation and hardship. They endured the most terrible of conditions. Eventually, after the winter, they managed to get back to Hut Point in a journey lasting some eight weeks and from there to Cape Evans. They arrived back after the search party had gone to look for Scott's party and their safe return was the only piece of good news that the search party had.

During the winter at the Cape Evans hut several of the men would give talks on different subjects. Ponting talked about his various travels and showed some photographic slides. Meares talked about his travels in China and the Tibetan border and what had happened to John Brooke in Lololand. The day after Meares' talk Scott wrote in his diary [29Aug 1911] that "the lecture was extraordinarily interesting". He added: "The spirit of the wanderer is in Meares' blood; he has no happiness but in the wild places of the earth. I have never met so extreme a type. Even now he is looking forward to getting away by himself to Hut Point, tired already of our scant measure of civilization".

There is no doubt that Meares got on well with the others. Wilson had nothing but praise for him. When the *Terra Nova* was sailing to Antarctica, Wilson wrote in his diary [21Dec1911] : "Meares is a capital chap and the amount of active life he has seen is extraordinary ... typically a man of action and a most entertaining messmate and full of fun". In the same entry he mentioned that many of the penguins they saw on the ice were attracted to the singing of some of the men and added: " Meares is the greatest attraction, and he has a full voice which is musical, but always very flat. he declares that 'God save the King' will always send them into the water and certainly it is often successful".

Later, in a letter to his wife, Wilson wrote that "Meares and Oates, the Dragoon, are just the finest men one could hope to meet".

Meares was particularly friendly with Oates. In the first published biography of Oates [L&C p129] the authors wrote : "with his curious, enigmatic life of

Dimitri and Meares at the blubber stove in the stable, preparing dog pemmican out of seal meat.

travelling about and exploring, his rejection of polite society, his scruffy unshaven appearance and his indignant individualism, Meares was very much a man after Oates' heart. They spent many an hour together at the blubber stove in the stable, preparing dog pemmican out of seal meat – they made about eight hundredweight of it". And both Meares and Oates were critical of Scott's view on transport and the various travel arrangements that he envisaged. On one occasion "Meares said that he reckoned Scott should buy a shilling book on transport. Scott overheard and was not pleased."

Of the different methods of travel, it was too early to pass any comment on the motor sledges. That would not be possible until the Southern Party got under way. Neither Oates nor Meares, though, were enthusiastic about the motors. Oates was reported [LPE p394] as saying "we both damned the motors. 3 motors at £1,000 each, 19 ponies at £5 each, 32 dogs at 30/- each. If Scott fails to get to the Pole he jolly well deserves it".

As far as the dogs were concerned Meares had no reason to fault their performance. Oates had tried his best with the ponies – and during the winter was to try and build up their strength – but already nearly half had died.

The hut at Cape Evans was divided into cubicles for the men. The Royal Navy petty officers and able seamen had their own space, separated from the officers and scientists by using the wooden packing cases as a dividing wall. Many of the officers and scientists had tried to make their cubicles as unique and comfortable as they could and each cubicle was given a nickname. Meares was with four others – Bowers, Cherry-Garrard, Oates and Atkinson. Their cubicle was plain and simple; it was nicknamed 'The Tenements'.

During the winter Cherry-Garrard typed fifty pages of what was to be another edition of 'The South Polar Times' : the earlier editions had been produced during Scott's *Discovery* expedition. This edition contained some drawings by Wilson and articles by various of the expedition members. Meares contributed an article on his travels into Lololand. The edition was bound by Day with venesta three ply boards from the packing cases, carved with the monogram 'SPT' and edged with grey sealskin. Later, in England, the different editions were published in facsimile by the publishers of Scott's diaries.

Scott held a church service every Sunday morning. Oates, though, refused to attend. In a number of letters to his mother he described his dislike of Scott and his mother was later to never forgive Scott for what she had understood to be his actions and what she thought was his unfair attitude to her son.

Inevitably there was some dissension among the men in the hut while living together at close quarters during the Antarctic winter but it is doubtful whether

TERRA NOVA 153

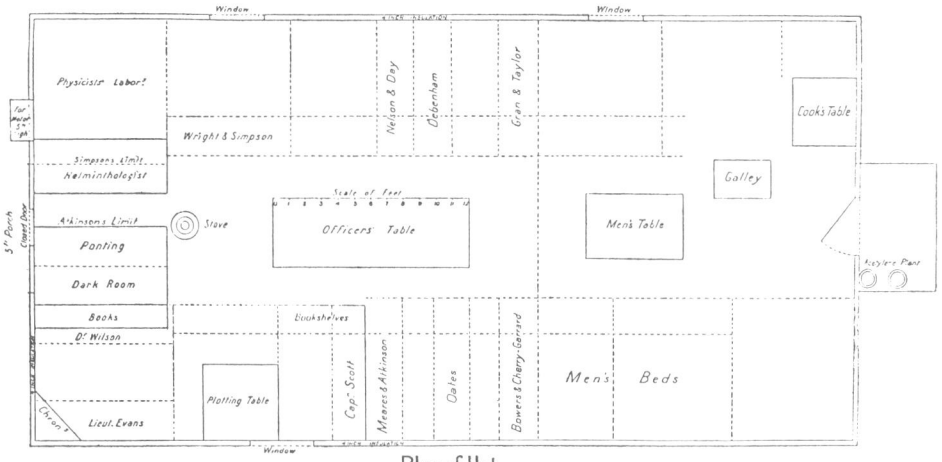

Scott's Hut Plan. Officers and Other Ranks carefully separated.

The Tenements, L to R, Cherry-Garard, Bowers, Oates, Meares, (top), Atkinson.

this was greater than that experienced by many other Antarctic expeditions. Frank Debenham wrote of Scott [FS p241 – "His temper is very uncertain ... in crises he acts very peculiarly ... what he decides is often enough the right thing I expect, but he loses all control of his tongue and makes us all feel wild ... I cannot say he is the least popular, still we are all prepared to follow him ... the marvellous part is that the Owner [Scott] is the single exception to a general sense of comradeship and jollity amongst all of us". Meares, however, was critical of Scott to the extent that he had made up his mind that, when the *Terra Nova* came back to Cape Evans at the end of the year, he would go back to New Zealand with it and not stay for another year in the Antarctic.

Not all, though, stayed in the hut throughout the winter. Wilson had wanted to look at the eggs of the Emperor penguins at Cape Crozier where there was a large colony. Wilson wanted to collect some fertile eggs and examine their embryos to try and trace their evolutionary history. If it was true that the birds' feathers had developed from the scales of reptiles then Wilson hoped this could be established by looking at the embryos under a microscope at several stages of development. He thought the penguin eggs would be at the right stage of incubation in July.

Wilson, Bowers and Cherry-Garrard set off for the one hundred and forty round trip to Cape Crozier and back on the twenty seventh of June. Their journey, in the most appalling of conditions – at one point the temperature dropped to minus 77 degrees fahrenheit – was to take five weeks. Cherry-Garrard called it 'the worst journey in the world'. His book of that name remains one of the best of all accounts of travel in the Antarctic. They brought three eggs back with them.

On the twenty fifth of August the sun reappeared. There was still much to do for the summer season and the journey south. George Simpson, the meteorologist, laid a telephone line of bare aluminium between Cape Evans and Hut Point. Scott, Bowers, Simpson and Edgar Evans man-hauled sledges to the mountains to the west of their base and up the Ferrar Glacier – a round journey of one hundred and fifty miles. A number of other small parties were sent out to dig out the depots at Safety and Corner Camps and take further supplies there. Meares twice took supplies out, with his dog team, to Corner Camp.

On the first occasion Scott wrote in his diary [17 Oct1911] : "Meares got back from Corner Camp at 8 a.m. Sunday morning – he got through on the telephone to report in the afternoon. He must have made the pace, which is promising for the dogs. Sixty geographical miles in two days and a night is good going – about as good as can be".

Meares made his second journey a week later and Scott wrote [24 Oct 1911]: "Meares this morning announced his return from Corner Camp, so that all stores are now out there. The run occupied the same time as the first, when the routine was : first day 17 miles out; second day 13 out, and 13 home; early third day run in. If only one could trust the dogs to keep going like this it would be splendid. On the whole things look hopeful".

Scott, however, had never intended that the dogs would be used on the journey beyond the Barrier – in spite of his complimentary remarks about their performance – and there is no indication whatsoever that he ever considered taking Meares on the final party to the South Pole. His idea was to use the motor sledges and the dogs to get out on the Barrier, use the ponies to get to the edge of the Barrier by the Beardmore Glacier and then for three teams of four men to man-haul the sledges. They would go up the Beardmore and then on to the polar plateau, about ten thousand feet above sea level. Two of the teams would turn back at different points, leaving the final party to get to the Pole. As far as that final party was concerned, however, there was considerable discussion and speculation among the entire party as to its composition.

On the thirteenth of September Scott outlined his plans in some detail. The two remaining motor sledges would leave Cape Evans on the twenty-fourth of October. Teddy Evans would lead the motor party with Bernard Day, the engineer, Bill Lashly, the Royal Navy chief stoker and Hooper, the former Royal Navy steward. They would leave Cape Evans first and, hopefully, wait for Scott's main party with the ponies some sixty miles beyond One Depot, at 80° 30' south. Scott and nine men would leave on the first of November, each walking with one of the ten remaining ponies. Meares and Dimitri Gerov would leave with the dogs afterwards.

In the event, the first of the motor sledges broke down just after Safety Camp and the second just after Corner Camp. Evans and the other three men then man-hauled their sledges to 80° 30' south to await Scott and the ponies.

He had intended to take the ponies the four hundred miles across the Barrier to the foot of the Beardmore Glacier. But the growing weakness of the ponies and the death of all of them before they reached the bottom of the Beardmore Glacier – and the slow speeds they made up till then, coupled with being held up by bad weather – meant that Scott had to change his plans. The dogs had performed well so he decided they should go on till just beyond the Lower Glacier Depot. It was longer than Scott had intended.

"The whole journey, there and back" [to and from the South Pole] said Scott when outlining his plans on the thirteenth of September [TQL

Debpp102-3] "is about 1,530 miles. Since Shackleton's figures are our best guide, I have laid these prospective plans upon data taken from his book … on this calculation the whole journey will be 144 days". Scott took with him a copy of Frank Wild's diary. Frank Wild, who had served on the *Discovery* expedition, had refused Scott's invitation to join the *Terra Nova* expedition. But Scott took a copy of Wild's record of the journey with Shackleton to the furthest south of 88° 23' south : it contained the mileages and positions of Shackleton's party for each day of their journey.

Scott then, a few days before the motor sledges started off, wrote out detailed letters to each of those responsible for different parties. He wrote one to Evans, in charge of the motor sledges. He wrote a long letter to Pennell [SS p153], the commander of the *Terra Nova*. In it he included a note about who was likely to want to return to New Zealand without waiting over for the second winter : Griffith Taylor whose leave of absence from the Commonwealth Weather Service in Australia Service would have expired, Herbert Ponting "who will have completed his work" and Anton Omelchenko "who has had enough of it" (though Scott did add that "he has worked like a Trojan and is an excellent little man".)

Scott also mentioned " Meares may possibly return; it depends on letters from home" – letters which the *Terra Nova* would bring down from New Zealand. Scott's comment was a reference to Meares possibly wishing to go back because of bad news on the health of his father. This may, in one way, have been true but it may also have been a polite way of disguising the real reason why Meares would want to return : he was getting disillusioned with the Scott expedition and certainly did not want to stay over for a second winter. But his father was ill and when Meares returned to New Zealand he got the news that his father had died. Oates, however made clear the reason for Meares wanting to go back with the *Terra Nova* in his last letter to his mother before Scott's party set off on the southern journey when he had written [SPRI ms1317 1-2] "Meares goes home in the ship. He is a very good chap although he did buy those rotten ponies. He told Scott he was going to clear out whatever happened. I don't think there is much love lost between them".

Scott also wrote to Simpson, whom he left in charge at Cape Evans, and to Griffith Taylor, another of the party's geologists who was to lead Frank Debenham, the Norwegian skier Trygve Gran and Robert Forde, one of the Royal Navy petty officers, to undertake geological exploration of the coast of Victoria Land. This was called the 'Western Party'. He also wrote to Meares.

Meares at the pianola, which also served as a bookshelf and a safe place for delicate instruments.

There has been much debate about exactly what instructions Meares did receive from Scott and to what extent the decision of Scott to take Meares and the dogs further than intended on the Barrier on the southern journey did actually affect the return of the South Pole party.

In his letter to him [20 Oct 1911] Scott stated that Meares should take some further supplies of oil cake and forage to Hut Point and to Corner Camp. Then he should go to One Ton Depot to meet Scott's party. "Under favourable conditions you should be back at Hut Point by December 19 at latest. After sufficient rest I should like you to transport to Hut Point such emergency stores as have not yet been sent from Cape Evans. At this time you should see that the *Discovery* hut is provisioned to support the Southern Party and yourself in the autumn in case the ship does not arrive."

Scott went on to say that during December "or early in January" Meares should make a second journey to One Ton Depot to leave further supplies there and that "this depot should be laid not later than January 19, in case of rapid return of first unit of Southern Party." Then – "supposing that you have returned to Hut Point by January 13" – Scott wanted him to go back to Cape Evans to assist in helping to unload the *Terra Nova*, which should have returned to Cape Evans by then.

Scott then laid down the need for another journey "About the first week of February I should like you to start your third journey to the south, the object being to hasten the return of the third Southern unit [this, Scott envisaged, would be the successful South Pole party] and give it a chance to catch the ship. The date of your departure must depend on news received from returning units [the supporting parties], the extent of the depot of dog food you have been able to leave at One Ton Camp, the state of the dogs etc."

The critical point about meeting the returning – and, hopefully successful – South Pole party was then spelt out by Scott "Assuming that the ship will have to leave the Sound [Cape Evans by McMurdo Sound] after the middle of March, it looks at present as though you should aim at meeting the returning party about March 1 in latitude 82 or 82.30 If you are then in a position to advance a few short marches or 'mark time' for five or six days on food bought, or ponies killed, you should have a good chance of effecting your object.

"You will carry with you beyond One Ton Camp one XS ration [basic food pack], including biscuit and one gallon of paraffin, and of course you will not wait beyond the time when you can safely return on back depots. You will of course understand that whilst the object of your third journey is important, that of the second is vital. At all hazards three XS units of provision must be got to One Ton Camp by the date named, and if the dogs are unable to perform this service, a man party must be organised".

In fact all these instructions to Meares were vitiated by Scott's decision to take Meares and the dog teams further than originally planned. Scott's original idea was that Meares and the dogs would turn back at about 81° 15' south; but he changed his mind and they went on to 83° 35' south, a few miles beyond the Lower Glacier Depot. This meant that Meares and the dogs went one hundred and forty miles further south than planned and also had a further one hundred and forty miles to go on the return journey to Cape Evans. This was to have a fundamental change to the proposals in Scott's letter of instruction – and also to the food available to Meares, Dimitri and the dogs.

When Evans and the motor sledge party started off on the twenty fourth of October, Amundsen – with four companions and fifty one dogs – was already on his way : they had set off from *Framheim* (the name they had given to their base by the Bay of Whales) for the Pole on the twentieth of October..

MOTOR BREAKDOWN

The second motor sledge broke down on the first of November. Evans and his three companions then man-hauled for two weeks and reached the appointed meeting place with Scott and the pony party at 80° 30' south on the fifteenth of November. They got there first : Evans had been determined not to be overtaken by the Scott party. In fact they waited for a week for Scott and the ponies to catch them up. But it had been hard work.

Meares and the dogs caught up with Scott on the seventh of November and went with the pony party until they got to the meeting with Evans and his party on the twenty first of November. They all went on together but three days later the first pony was shot. Scott decided that Day and Hooper from Evans' team should go back to Cape Evans. Teddy Evans and Lashly would join the others and, with the others, would help form the three groups of four men who would – after all the ponies were shot – man-haul the sledges. It would mean that Evans and Lashly would have man-hauled for several hundred more miles than the others and it would make their return as the last supporting party all the more remarkable.

As the party went forward they laid depots about every sixty-five miles for the use of the returning supporting parties and the final party. The dogs were going well. They started after the pony party each day of travel, partly so they could cut up the dead ponies and store them in food depots. They then soon caught up with the ponies. On the fourth of December the whole party was held up by a blizzard which lasted for four days. Then they were near the Beardmore Glacier.

The Beardmore was some one hundred and twenty miles long and, in places, forty miles wide. Its ascent was over nine thousand feet. Shackleton's party had shown the way up in December 1908. They placed a depot at the foot – named the Lower Glacier Depot – and a few miles southward Scott decided that Meares and the dogs should go back. They had more than done their job : in fact their carrying of food for two weeks longer than planned, and thus relieving the others of some of the weight of the food, had undoubtedly helped the rest to make up for time and stand a chance of the final party making it to the South Pole.

There has been comment that Meares took from some of the depots food that was intended for the supporting parties, but not him or Dimitri. In reality they had to take some food because they were in fact away for nearly four weeks longer than anticipated. And their journey back was fraught with difficulty.

In his famous book 'The Worst Journey in the World' Cherry-Garrard wrote about Meares' return journey [WJW pp382-3] – in fact, the only account of any detail in any of the books about the whole expedition.

All the three parties now going forward up the Beardmore Glacier and then on to the polar plateau "owed a great deal to Meares, who, on his return with the two dog teams, built up the cairns which had been obliterated by the big blizzard of December 5 - 8. The ponies' walls were drifted level with the surface, and Meares himself had an anxious time finding his way home. The dog tracks also helped us a good deal : the dogs were sinking deeply and making heavy weather of it."

Cherry-Garrard, Atkinson, Charles 'Silas' Wright (the young Canadian physicist), and Keohane were the first supporting party to be sent back – from 85° 5' south, near the top of the Beardmore Glacier. They started back on the twenty second of December. Prior to their returning Scott instructed Atkinson to bring the dogs out to meet the returning parties if Meares had gone back to New Zealand on the *Terra Nova*.

Cherry-Garrard went on : "At the Barrier Depots we found rather despondent notes from Meares about his progress. To the Southern Barrier Depot he had uncomfortably high temperatures and a very soft surface, and found the cairns drifted up and hard to see. At the Middle Barrier Depot we found a note from him dated December 20. Thick weather and blizzards had delayed him, and once he had got right off the tracks and had been out from his camp hunting for them. They were quite well : a little eye strain from searching for cairns. He was taking a little butter from each bag [of the three depoted weekly units], and with this would have enough to the next depot on short rations.

At the Mount Hooper depot " the news from Meares was dated Christmas Eve, in the evening : the dogs were going slowly but steadily in the very soft stuff, especially his last two days. He was running short of food, having only biscuit crumbs, tea, some cornflour, and a half cup of pemmican. He was therefore taking fifty biscuits, and a day's provisions for two men from each of our units."

Far from Meares taking out unnecessary food from the depots, he only took a very bare minimum – and he had had no alternative in view of his longer journey out and back. Cherry-Garrard continued : "The dogs had the ponies on which to feed : to make up the deficiency of man-food we went one biscuit a day short when going up the Beardmore : but the dogs went back slower than was estimated and his provisions were insufficient".

Cherry-Garrard realised that Meares would be very unlikely to get back to Cape Evans and then return to One Ton Depot with more supplies for the returning parties and "it was uncertain whether a man-hauling party with such of this food as they could drag would arrive at the depot before us". However, he and his three companions, were delighted to see when arriving at One Ton Depot on the fifteenth of January that a man-hauling party had been there with further supplies.

Day and Hooper had arrived back at Cape Evans on the twenty first of December. Five days later they had gone – with Edward Nelson, the biologist, and Thomas Clissold, the cook – to One Ton Depot. Cherry-Garrard and the other three men then returned successfully to Cape Evans on the twenty sixth of January.

Meares and Dimitri Gerov had arrived at Cape Evans on the fourth of January. The original timetable laid out in Scott's letter of instruction to Meares had now been overtaken. The *Terra Nova* came back to Cape Evans in early February and Meares and a dog team helped unload stores, letters and the seven Indian mules and a further eleven dogs. One of the letters was to Simpson from the Indian Meteorological Service saying they wanted him to return to his post in Simla in India as soon as possible.

Atkinson then, in accordance with Scott's instructions to him when his supporting party left Scott at the top of the Beardmore Glacier, took Dimitri and two dogs teams from Cape Evans to Hut Point. They left on the thirteenth of February. At Hut Point they were preparing their sledges and equipment for the journey south to meet Scott and his returning – and hopefully, successful – party. Bad weather delayed their departure and then Tom Crean walked into the hut.

The last supporting party of Scott's party had turned round to head north on the fifth of January 1912. As they waved goodbye to Scott's party they did not know that they would be the last people to see them alive.

The day before, Scott had decided that his tent party – Wilson, Oates, petty officer Evans and himself would be the party to go to the South Pole. He, surprisingly, decided that he would also take Bowers with him. That meant that the five men would have to share the four man tent and it also meant difficulties with the food bags they carried – and those at the depots – which had been allocated into four man bags. But Scott reckoned that Bowers would make all the difference to their successful attainment of the Pole.

This meant that Teddy Evans and the two other RN petty officers, Bill Lashly and Tom Crean, would be a three man party to return. They left Scott's party at 87° 34' south. They had to man-haul their way back for over six hundred miles in extremely difficult conditions. And Evans and Lashly had already travelled nearly four hundred miles more than the others by man-hauling from where the motor sledges had broken down, near Safety and Corner Camps, to the foot of the Beardmore Glacier. It would be a very close call.

Evans collapsed by the time they got back to One Ton Depot and Crean and Lashly pulled him on the sledge for the next six days. Then just before Corner Camp on the eighteenth of February, Lashly stayed with Evans in their tent while Crean decided to walk to Hut Point to try and get help. It was a thirty-five mile journey. Crean had no tent and only a few biscuits and two pieces of chocolate with him. In eighteen hours he got to Hut Point – a remarkable feat. There he met Atkinson.

Atkinson and Dimitri had got to Hut Point with two dog teams from Cape Evans in order to start south to meet Scott's returning party, but now Crean's news changed that. They took the dogs, found Lashly and Evans and took them back to Hut Point – travelling the thirty-five miles in just over three hours : a striking example of how quickly the dog teams could go.

SCURVY

It was clear then to Atkinson that Evans was in serious danger of dying and reckoned that, without treatment, he would die in a couple of days. Evans was suffering badly from scurvy, though, oddly enough, Crean and Lashly showed no signs of it. Atkinson then decided that he should get Evans back to Cape Evans and nurse him back to health. For their efforts in looking after Evans and getting back to Hut Point, Crean and Lashly were – on return to England – awarded the Albert Medal (later merged with the George Cross) by King George V.

Because Meares had arrived back at Cape Evans later than originally planned and was going back with the *Terra Nova*, there was no way in which he could now lead the dog teams out to try and meet Scott's returning party at One Ton Depot, let alone the latitude s of 81° or 82° as envisaged by Scott in his letter of instruction. Simpson and Taylor were also returning on the *Terra Nova* and therefore Atkinson was the senior man left at Cape Evans – and the only doctor. He decided he now had to stay with Evans and would have to abandon his idea of leading the dogs to meet Scott's party. The problem then was who was going with Dimitri Gerov to take the dogs back to try and meet Scott?

Atkinson sent Dimitri back to Cape Evans with a message saying that either Wright or Cherry-Garrard would now have to go with Dimitri and the dogs to search for Scott's party. But Silas Wright was the only scientist left at the base and he had to remain in charge of all the scientific work and observations there. That left Cherry-Garrard to take the dogs back on to the Barrier with Dimitri. He, however, had little experience of driving a dog team; but there was no one else.

He and Dimitri arrived at One Ton Depot on the second of March. There they waited one week but there was no sign of Scott's party. A storm blew for several days. Cherry-Garrard had been told by Atkinson that dogs would be required for another Antarctic season and was not keen to risk them by going further south than One Ton Depot. Scott had originally said – when asking in a letter to his agent in New Zealand, Sir Joseph Kinsey, for fourteen more dogs to be sent down south, as well as the mules – that if he failed to reach the South Pole in 1911 - 12 then he might well make a further attempt in the following season and for that he would need more dogs as well as the mules [AT ms0022-6]

After six days at One Ton Depot Cherry-Garrard decided that he had no option but to return. He had no idea what had happened to Scott's party or where they were. The weather was atrocious. He knew that Atkinson's instructions to him had been to travel to One Ton Depot as soon as possible and that if Scott had not arrived before him "I was to judge what to do"

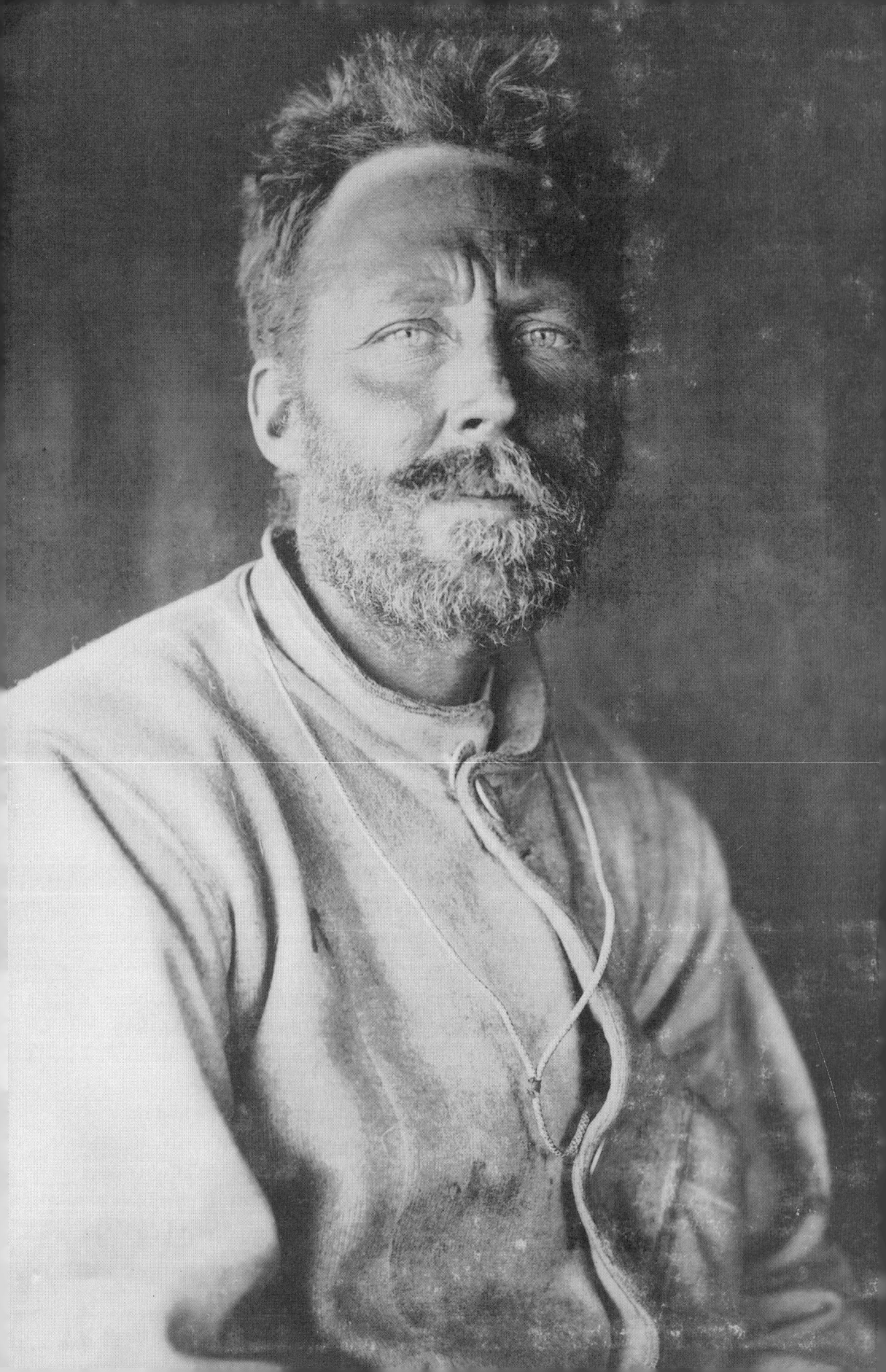

[WJW p416]. Furthermore he recorded later Atkinson had told him "That Scott was not in any way dependent on the dogs for his return" and "That Scott had given particular instructions that the dogs were not to be risked in view of the sledging plans for next season".

Cherry-Garrard's decision was to haunt him for the rest of his life. If he and Dimitri had gone further south, they might have met Scott's party and rescued them – but how were they to know?

The *Terra Nova* left Cape Evans on the fifth of March 1912. A replacement cook, Archer, and another Royal Navy petty officer, Thomas Williamson, were landed to join the Cape Evans party. On board the ship to go back to New Zealand, were Taylor, Simpson, Clissold, Forde, Day, Ponting, the very sick Teddy Evans and Meares.

Scott's party meanwhile was experiencing conditions of extreme cold and very difficult travelling surfaces on their journey. Scott reckoned that they had much worse weather conditions than Shackleton's party three years earlier. They had reached the South Pole on the seventeenth of January 1912. Amundsen had reached the Pole on the fourteenth of December. Amundsen's return was fairly straightforward. Scott's was over one month later and the weather was considerably worse and the cold was exceptional.

Amundsen's time in getting to the Pole and back was ninety nine days. They had left Framheim on the twentieth of October 1911 and arrived back – with eleven out of the fifty two dogs still alive – on the twenty-fifth of January 1912. The *Fram*, which had spent the Antarctic winter in Buenos Aires, had returned to the base and five days after Amundsen's party had arrived back, sailed for Tasmania with all safely on board. On reaching Hobart Amundsen telegraphed his news.

Scott had left Cape Evans on the first of November 1911. On the twenty first of March 1912 he, Wilson and Bowers had made their final camp – just eleven miles south of One Ton Depot. On the seventeenth of February petty officer Evans had died. On the sixteenth of March Oates and left the tent and walked to his death. On the twenty first of March a fierce blizzard prevented Scott and his two remaining men from leaving their tent. Their food supplies and fuel were pitifully low. They had taken one hundred and forty-one days from starting off to where they made their last camp. On the twenty ninth of March Scott made his last entry in his diary.

After a second winter in the Cape Evans hut, a relief party set off on the thirtieth of October. There were eight men with the seven Indian mules. The next day two dogs teams set out with Cherry-Garrard, Dimitri and

Meares upon return from the Beardmore Glacier. Jan, 1912.

Atkinson. This time the mules seemed to do much better than the ponies. The whole party reached One Ton Depot on the tenth of November and two days later found Scott's tent. There they found the three bodies and Scott's diary. They went twenty miles further south to look for Oates body but failed to find it – though they did find Oates' sleeping bag which the others had left after Oates had walked from the tent. That sleeping bag is now with the Scott Polar Research Institute in Cambridge.

On the twenty fifth of November Cherry-Garrard, Dimitri and Atkinson with the dogs – having pushed ahead of the mules and the rest of the oarty – reached Hut Point. There they found a letter inside telling them that Victor Campbell's Northern Party had reached Hut Point on the sixth of November and were now back at Cape Evans.

On the eighteenth of January 1913 the *Terra Nova* arrived back at Cape Evans with a now recovered Teddy Evans on board. On the twenty sixth of January, with the whole shore party – apart from Scott, Wilson, Bowers, Oates and Edgar Evans – safely on board, the *Terra Nova* sailed away and arrived back in New Zealand on the twelfth of February. From there Atkinson telegraphed the news.

ARMY, NAVY & AIR FORCE

ARMY, NAVY & AIR FORCE

After a short stay in New Zealand Meares arrived back in England, with Bernard Day, the motor engineer, on the tenth of July 1912. They landed at Plymouth. Then, for a while, Meares shared a flat with Herbert Ponting near Oxford Circus in London. He had to sort out his father's estate and it is clear that the income from the estate meant that he did not have to search for a regular job.

He retained his interest in the Antarctic and over the years was to meet up with several of the members of the *Terra Nova* expedition. When back in England he received a letter from Dimitri Gerov. Gerov had dictated it on the second of January 1913 to Debenham and it was sent to Meares when the *Terra Nova* arrived back in New Zealand.

In it Dimitri had written about the state of the dogs and then the details of the search party for Scott. He wrote that "the mules were pulling rather more than ponies did last year". Two dog teams were taken, including seven of the new dogs brought down by the *Terra Nova*. He continued "on the Barrier your team worked pretty well". He also mentioned that a party of six men – including Priestley, Debenham and Trygve Gran, – had gone to Mount Erebus and that four of them had climbed to the top.

At the end of the letter Debenham had added a few words: "Just a line at the end of this to say that Dimitri has looked after my room (the dark room) like a brick all the year and has done all my dirty work for me. He's a rattling good chap".

In the middle of 1913 Meares was asked by Teddy Evans to take charge of the Antarctic section in the Army and Navy exhibition in Earl's Court. He was also present at the showing of Ponting's film of the Scott expedition to King George V and Queen Mary at Buckingham Palace. During the early part of the year he also did some lecturing but there is no record as to where he did this.

In 1914 Meares attended – along with Edward Atkinson – the unveiling of a brass memorial plaque to Oates in the north wall of Gestingthorp Church in Essex. There were several other memorials to Oates including one at Eton (where he had been to school) and the Cavalry Club. Major-General Allenby, who had commanded the Sixth Iniskilling Dragoons for a while during the Boer War, gave the address. Ever after, until her death, Oates' mother would polish the memorial every week.

Then, at the beginning of August 1914 the First World War started. On the first of August Germany declared war against Russia and two days later against France. On the fourth of August Britain declared war against Germany. The next day Austria-Hungary declared war against Russia and one week later Britain and France declared war against Austria-Hungary. General mobilization began in each of the warring countries.

THE ARMY

Meares joined the Northumberland Hussars , part of the Seventh Division of the British Army, as a commissioned officer. Many in the Division were soldiers who had served in India and also the Boer War. The Division was part of Fourth Corps under the command of Lieutenant-General Sir Henry Rawlinson. In the third week of September 1914 Meares was in the Flanders region of Belgium with his regiment and was to become heavily involved in what became known as the First Battle of Ypres.

The British Army at the time was a small regular army but territorial units were called up as soon as hostilities started. The Battle of Ypres started on the twentieth of October 1914 when the superior forces of the German Army , under General von Falkenhayn, began their Flanders offensive. It ended on the twenty-second of November 1914. All those serving under the command of General Sir John French with the British Expeditionary Force – as the British Army units in France and Belgium were called – were called the 'Old Contemptibles'.

Meares – like all those who served in First Ypres (as the battle was called) – was therefore an 'Old Contemptible'. The description came from the famous 'Order of the Day' given by the German Kaiser Wilhelm II on the nineteenth of August 1914 : "It is my Royal and Imperial Command that you concentrate your energies, for the immediate present upon one single purpose, and that is that you address all your skill and all the valour of my soldiers to exterminate first the treacherous English; walk over General French's contemptible little Army".

Meares served with the British armed services from the beginning of the war until 1919. He was one of the few men who held commissioned rank in the Army, the Royal Navy (through serving with the Royal Naval Air Service) and the Royal Air Force. He was demobilised in May 1919 from the newly created Royal Air Force with the rank of Lieutenant- Colonel.

MARRIAGE

At the outbreak of war Meares was thirty seven years old. Prior to joining the Army he had met Annie Christina Spengler. She was born on the twenty ninth of December 1891 and was fourteen years younger than him. She was about five feet and four inches tall and had brown eyes and brown hair and used the name 'Christine' rather than 'Christina'. It is clear from the letters he wrote to her from Belgium and France in the early weeks of the war that he was deeply in love with her. When back home on leave he married Christine on the sixth of February 1915. His letters were eventually left to the British Columbia Archives in Canada by his wife and deposited there after her death – some thirty seven years after his.

The letters Meares wrote had to be passed by the official War Office censor, though later when he was with the Royal Navy Air Service, he was himself the official censor at his unit. Because of the censorship Meares could not give details of where he was stationed or anything else of a military nature but he was able to receive letters and parcels that Christine sent him.

In one of his first letters he wrote [BC ms0455]: "We started from Southampton to a certain new port where we landed and went inland to another larger town where I had a busy time arranging quarters for the horses and men as about 20,000 men were in the town. Next day we returned to another large place on the sea which is simply filled with refugees wounded and troops of all kinds. Tomorrow we expect to go off to a very hot corner. We were to go off this morning but the trains are disorganised and we may have to go off any time in the middle of the night."

When the war started the German forces made rapid advance into Belgium and France and within a month were only thirty miles from Paris. Then they were held by the British and French forces at the Battle of the Marne and retreated. Each side then began what was known as 'the race to the sea' : to try and outflank the other side. Particularly the Germans, having captured Antwerp, wanted to cut through Ypres and capture the French ports of Calais, Dunkerque and Boulogne. If they had succeeded the outlook for the British forces would have been grim. Between the eighth and nineteenth of October the British Expeditionary Force went to Ypres to bolster the French and Belgian armies in the area.

The BEF consisted of seven infantry divisions and three cavalry divisions. Divisions of Indian troops were on their way to reinforce them and they would prove a valuable addition to French's Army. Against them were many more German units. One estimate was that, at Ypres, Hamilton's Seventh Division was outnumbered by up to eight times the number of German soldiers.

During a lull in the fighting Meares wrote to Christine "I am writing to you in the drawing room of a beautiful chateau and the horses are all on the lawn in front of the big 4.7 [sic] guns are making a great din all round. At intervals the Germans drop big shells on this house which is very annoying. We have had a very hot time here, lately the 7th Division has been holding up a whole German Army Corps for about 5 days and has had a rough time but now other troops and guns are coming up to help, so I hope it will be."

He went on : "Yesterday the regiment was in a very hot corner indeed but the men were splendid and we were very fortunate, not a man killed but some wounded; the shrapnel fire was very heavy and lots of very big shells. We were making some soup in a farm when 5 big guns dropped around it, but we and all the horses got away."

He then described an incident affecting the major in his unit : "My major with whom I mess was shot through the lungs. I did not think he would pull thro, as we could not get a stretcher or ambulance to take him away. After a long time we got a motor ambulance but it was wrecked about a mile away. The doctor and I stayed with him till night and got him into a motor car and away just as the shells were beginning to fall again. I was very tired as I had no coat and had no grub for about 24 hours."

He added : "It is difficult to realise how bad this war is till one sees the thousands of refugees and hundreds of burning houses all round" and then "You might send this letter to Ponting to read, 47 Oxford Mansions. I am too tired and will have to go out again soon".

Despite their superior numbers the German units were unable to break through the British line and although the town of Ypres was severely damaged it was not captured : nor was it ever captured throughout the war. Meares was proud of his unit and wrote to Christine in one letter "this is the finest crowd of infantry and artillery I have seen" and, again, in another letter "This [is] supposed to be the finest infantry division in the Army – all old soldiers from India and abroad".

In another letter Meares wrote; "Many thanks for your sweet letter just received. I was very glad to get it. It really seems years since I left you [the letter was written in the third week of October]. We have had a very strenuous time, marching all the time and some rather tight places. Now we are in a certain town and have been out hunting Uhlans [German light cavalry] this morning. We shot one and I had a shot at some, some fifty yards in a thick wood, but did not get them. They seem to be very bad shots.

"The country here is very beautiful wooded country and very pretty villages, the people are very glad to see us and bring out fruit, coffee etc. and think that we will help them, but I am sorry to say we have not been able to do much for them. This is a beautiful town and is full of troops of all kinds. I expect that there will be a big fight here soon, of course I can't say where we are or have been but you will be surprised when you do hear."

Already at that point of the war, the use of aeroplanes by both sides was starting to become common. In his letter Meares wrote "We have shot down 3 of their aeroplanes already and yesterday one of our aeroplanes was chasing theirs but I did not hear whether they got it." He added "It seems to me that the war will last a long time".

At the end of October Meares wrote : "Since I wrote last we have had a terrible time. A number of officers have been wounded and gone home but this regiment has been very fortunate compared with the others. At present we are having a few days rest, but even now the guns are going hard all round and shrapnel is bursting fairly near. Fortunately crowds of soldiers have arrived here and the place is full of English and French troops so I hope things will not be so strenuous. This regiment has done splendid work. There is so much that I would like to tell you but cannot do so. I hope that I will get back some day and tell you about it. No one can have any idea how terrible this war is and this seems to be the hottest corner."

By the twenty second of November the fighting had reached stalemate. Neither side had been able to outflank the other. There now stretched a line of trenches for some four hundred miles – from the North Sea to Switzerland. There had been a high price to pay. The total loss to the British Expeditionary Force from the fourth of August to the thirtieth of November 1914, in terms of men killed, wounded or taken prisoner, was eighty-six thousand men. British casualties in the Ypres area from the fourteenth of October to the thirtieth of November alone were forty two thousand men. Even by the fifth of November the Seventh Division had gone from twelve thousand men to a little over three thousand. The casualties were enormous.

Earlier in November Meares had written : "We have come back from the firing line for a few days rest. There are very few of the splendid 7th Division, the fighting has [been] beyond anything in the history of the world, and we have had to bear the brunt of it. Lately the Germans have been pouring hundreds of shells into the town, huge shells 16 inches across, which they used at Antwerp and they make holes in the ground big enough to hold a house, and even during the night they drop close all round us, it gets on ones nerves and we get very little sleep. It is fine to be out of reach if only for a day or two."

A few days later he wrote : "We are still in rest camp having an easy time but I expect we will soon be back in the thick of it. I only hope they don't send us back where we were before, it was too hot for words. I hear that the Prussian guard tried to get through just after we left and was cut to pieces. I hear that a lot of territorials are coming out now which will give us an easier time."

The next day he wrote : "Just a line to say we are off again tomorrow to the fighting line, but I think not to the same place where we were before, that is too hot; the Germans have been trying to take it ever since we left … this afternoon General French arrived here suddenly in the rain, and congratulated the regiment on the splendid work they had done under very difficult circumstances, and made a very flattering and rousing speech".

HEART BREAKING WORK

On the nineteenth of November Meares wrote : "We are back in the fighting line but in a rather quiet place and there are lots of troops, so we don't have much to do at present … I am afraid the war will not finish by this Xmas but perhaps I might be back by the spring time if all goes well. I have had very unpleasant work lately, turning all the French farm people out of their houses behind the trenches. It is heart breaking work but necessary as people in the pay of the Germans have been cutting our wires and shooting the soldiers, so the only thing is to clear all civilians from the fighting area. It is beginning to get cold now, frost ice and snow, getting up at 5 am is rather chilly work".

On the twenty sixth of November he sent to Christine two messages to the troops from General Sir Douglas Haig and Lieutenant -General Sir Henry Rawlinson, specifically to the Seventh Division, saying "please keep them carefully". Rawlinson had written [23Nov1914] : "I desire to place on record my own high appreciation of the endurance and fine soldierly qualities exhibited by all ranks of the 7th Division from the time of their landing in Belgium. You have been called to take a conspicuous part in one of the severest struggles in the history of the War, and you have had the honour and distinction of contributing in no small measure to the success of our arms and the defeat of the enemy's plans. The task which fell to your share inevitably involved heavy losses but you have, at any rate, the satisfaction of knowing that the losses you have inflicted upon the enemy have been far heavier. The 7th Division have gained for themselves a reputation for stubborn valour and endurance in defence, and I am certain that you will only add to your laurels when the opportunity for advancing to the attack is given you".

The First battle of Ypres was one of the most significant battles of the war. Meares had been part of it. Shortly after he was given some leave and returned to England. On the seventh of December he wrote to Christine, having arrived back in Belgium after his short leave. He wrote : "just a few lines to say that I got back all right and had a good journey. It was rather rough crossing the channel but I had a good lunch and we reached Boulogne at 5 pm. At 6 we started off in a motor bus and reached Merveille at 12 midnight. There we found a motor car which brought us and landed me here at 1.30 am. Things here are just the same as when I left; but they have had terrible storms of wind and rain and tonight is one of the worst nights I have known … the King came here and the regiment acted as guard of honour; the King went near the trenches and the Staff were very worried as he insisted on going along a road which is constantly shelled. I don't know what I will do as the colonel has written to the War Office saying that he wants me to stay on here".

While on leave Meares must have decided to apply for a transfer and it was to the Royal Naval Air Service.

ROYAL NAVAL AIR SERVICE

Air power was very much an unknown quantity at the beginning of the war; in fact Britain was lagging behind both France and Germany at the outset – in terms of pilots and aircraft. The introduction in 1909, by French engineers, of the rotary engine as the means of propulsion was a big step forward and pressure for a separate structure for the use of military aircraft started. There was, however, opposition. In 1910 the British Chief of the Imperial General Staff decried the whole notion of military aviation as "a useless and expensive fad" [RFC RB] and the First Sea Lord reckoned that "the naval requirement for aircraft was two".

However, pressure increased and the example of flying box kite aeroplanes in India in 1910 greatly impressed the Chief of General Staff in India, Sir Douglas Haig, in their reconnaissance role. Then, in February 1911, an army order was made for the establishment on an Air Battalion of the Royal Engineers on the first of April. Pressure for more expansion increased : in 1911 it was noted that the French had already two hundred aircraft in service while the new Air Battalion of the British Army had twelve. A sub-committee of the Committee for Imperial Defence looked at the issue and recommended the formation of a new flying corps, divided into two wings – one naval and one military.

The Royal Flying Corps was then formed from April 1912. In fact the Admiralty now wanted to pursue their own separate service. They already had their own training centre and finally the Royal Naval Air Service was recognised as a separate organisation in July 1914.

In August 1914 the strength of the whole of the Royal Flying Corps (including the Royal Navy aeroplanes) was just over two thousand officers and men with one hundred and thirteen aeroplanes and sea planes and six airships. By the end of the war in November 1918 the newly formed Royal Air Force had nearly three hundred thousand officers and men, with just under twenty-three thousand aircraft in one hundred and eighty-eight squadrons. The RAF was then the largest air force in the world.

Meares joined the RNAS and was stationed with number 4 Wing in France. Its main duty was reconnaissance and especially tracking and attacking – if possible – the German U-boat submarines and also attacking (by physically dropping bombs and direct rifle fire) the German occupied ports in Belgium. Meares' first certificate from the Wing Commander covered the period from the thirty first of March 1915 to the sixteenth of January 1917 . He was then Lieutenant-Commander Meares of the Royal Navy Volunteer Reserve. The Wing Commander wrote that during the period Meares had conducted himself "with zeal and to my entire satisfaction. Has acted as Intelligence and Transport Officer in both of which [roles] he has shown his competence. An excellent interpreter."

A later certificate covering the period between the twenty first of November 1917 to the first of April 1918 stated again that the same Wing Commander thought he had conducted himself "with energy, zeal and to my entire satisfaction". By then Meares had been promoted to the rank of Commander – the equivalent of Lieutenant-Colonel in the Army.

ROYAL AIR FORCE

In April 1918 Meares was transferred to the new Royal Air Force. This included both the Royal Flying Corps and the Royal Naval Air Service. Although new titles were given to the ranks in the RAF several officers still retained their Army titles. Surprisingly, as Meares had been listed as a Commander in the RNAS (where paradoxically his commanding officer had been titled as a

Wing Commander) he reverted to being a Lieutenant-Colonel in the RAF. A few years later the Royal Navy then established its own aircraft force – the Fleet Air Arm.

During the war Meares met some members of the *Terra Nova* expedition. Charles (Silas) Wright served throughout the war and finished as a Major in the Royal Engineers. He also was awarded the Military Cross. In a letter to Cherry-Garrard in July 1916 he mentioned that he "saw Meares a little while ago" [Si p379].

Earlier, Meares in a letter to his wife, postmarked the third of May 1916 and with an address of No. 4 Wing, RNAS c/o Naval Mail Officer, Dover, wrote "I dined with Commander [Teddy] and Mrs Evans. Evans is stationed here in Dunkirk and has several destroyers under his command. Campbell is also coming here in command of the *Mohawk*".

After the end of the war in November 1918 Meares was posted to the headquarters of number 22 Group of the RAF at Stirling in Scotland. RAF records [RAF MH 19 Dec 20] recorded that Meares was put on the unemployment list by May 1919 and "the fact that his name had gone from the list by March 1920 suggests that he had left the service altogether by that time, probably because of his age". In March 1920 Meares was forty three years old: but he was to do one more thing for the development of air power. He joined the British Aviation Mission to Japan.

JAPAN AND CANADA

JAPAN AND CANADA

By the end of the First World War Britain had built up the largest air force in the world. Its expansion since the formation of the Royal Flying Corps in 1912 had been enormous. The United States of America had also developed a large air force and was soon to overtake the Royal Air Force in size. Japan, which already had a large army and navy, wanted to increase the size and strength of its own air force.

During the war, Japan – an ally of Britain – had played very little active part in the conflict. The Japanese armaments industry had supplied large quantities of weapons and shells and Japanese ships had escorted troopships of Australian and New Zealand forces across the Indian Ocean. Japan's armed forces had also taken possession of the German colonies in the Pacific . But that was really the extent of their active involvement. The League of Nations had acquiesced in the Japanese occupation of the former German islands but, when Japan – in defiance of the League's wishes – started to fortify them in 1921, there was nothing the League was able to do about it.

The year after the war ended the Japanese Army had asked the French Government to assist in the training and development of their own army air force but, while the French agreed and sent a team of sixty one men, they had also taken the opportunity to sell a lot of old and outdated aircraft to Japan and their collaboration had not been very productive. The Imperial Japanese Navy then asked Britain to help with the training and development of its naval air service.

With hindsight, it was extremely dangerous for Britain to assist the Japanese – and the consequences for Britain and the rest of the world would be horrendous – but at the time there were ambivalent attitudes to the Japanese request. Japan was an ally of Britain and the alliance treaty was now nearly twenty years old. It was argued that if Britain did not help the Japanese then some other European power would do so. Furthermore, there could be some material advantages for the British domestic aircraft industry in dealing with Japan.

The United States of America was very unhappy about any British assistance to Japan and were particularly concerned about the implications of such assistance on their own relationships with countries of the Pacific and their military bases in the Philippine Islands.

The British Colonial Office, the Admiralty, the Air Ministry and the Foreign Office were all apprehensive too – with the Admiralty flatly opposed to any British involvement with a developing Japanese naval air service.

THE MISSION TO JAPAN

In the end it was agreed that a British Air Mission could go ahead – after all, why should Britain not help an allied nation? – but that it would be politic for it to be designated as a civilian mission. In fact of the thirty men appointed to the Mission, most were either serving or recently retired service personnel. Although the pretence of a civilian mission was maintained, and all those involved did not wear official RAF uniform, they did all wear special uniforms and were each given Japanese naval rank.

The Mission was composed of instructors, not only in flying, but in armament, technical design and administration. The effect was to improve the training methods, the structure and organisation of the Imperial Japanese Navy's air force – especially with the training of pilots in some of the newer types of aeroplane now being produced, and in the landing on and taking off from ships. The active British serving officers then returned to their own military units after the Mission's conclusion.

The Japanese Government had approached the Master of Sempill – Captain the Honourable WW Forbes Sempill, holder of a fifteenth century Barony and a keen aviator (and holder of an Air Force Cross) – to lead the proposed Mission. Meares was one of the first to be appointed and was the first to arrive in Japan at the commencement of the Mission in March 1921 while the others landed in Japan in the following month. Most of the others had agreed to serve for twelve months from April 1921 but the majority extended their annual terms until the Mission ended in 1924. In fact the Mission was effectively cut short by the disastrous earthquake that erupted in Japan in September 1923 but officially it is still known as the Mission of 1921 to 1924. As it happened, Sempill left Japan in October 1922 and Major HG Brackley then took over command of the Mission [OMRS]. Meares, though, returned home in November 1921.

It is not known how Meares came to join the Mission or if someone recommended him. He had been present at the Russo-Japanese War of 1904 to 1905 and had made no secret of this. He had performed well both with the Royal Naval Air Service and then the Royal Air Force during the war. He

180 MEN OF ICE

The British Aviation Mission to Japan, 1921
Meares and Orde-Lees are seated on the front row.
© Medals Research Society.

was not a pilot but had considerable experience of organisation and administration of an air force, its training stations, operational squadrons and the usefulness of aerial reconnaissance and attack.

Most of the official papers relating to what was called the British Air Mission – also called the 'British Civil Aviation Mission to Japan' – were not released by the British Government for up to fifty years and there is still some mystery about aspects of it. The names, though, of all of the members of the Mission have been published. One of the members was Thomas Orde-Lees.

ORDE-LEES

Orde-Lees had been commissioned into the Royal Marines in 1896 and had served in a warship off China during the Boxer Rebellion of 1900. He had joined Sir Ernest Shackleton's *Endurance* expedition to Antarctica in 1914 and had been given permission by the Admiralty – and specifically by Winston Churchill as First Lord of the Admiralty – to take leave of absence for this. He was originally taken on to be in charge of the aero-sledges which Shackleton hoped to use.

Shackleton proposed to make the first crossing of the Antarctic from the Weddell Sea to the Ross Sea, via the South Pole. In the event the *Endurance* never landed the shore party on the continent of Antarctica and the aero-sledges were not used at all. It is doubtful anyway whether they would have been of any use. Orde-Lees was then designated the expedition's storekeeper.

Orde-Lees was one of the early people to be appointed to the Aviation Mission. He had served with Captain Forbes Sempill, the Master of Sempill, on the Parachute Committee set up by the Air Board and Sempill specifically asked for him. Altogether some six thousand airmen were killed during the First World War and a great many were pilots. It was noted that later in the war the German air force did provide their pilots with parachutes. The pressure for their use in the Royal Air Force was growing.

It is surprising now to look back at the controversy which the use of parachutes in the Royal Air Force was to engender. The first descent from an aeroplane by parachute was made in 1914 just before the outbreak of war. Orde-Lees was one of those who could see obvious advantages for their use. But there were powerful opponents. Some senior officers argued that the availability of parachutes in aeroplanes could encourage pilots to jump prematurely or even avoid combat altogether. One dubbed it 'unmanly' [EIB p289]. After much discussion the Air Board did establish a Parachute Committee – with Orde-Lees as its secretary (he was awarded an Air Force

Cross for his role in this) – and it did recommend the use of parachutes in military aeroplanes. Its recommendation was, however, not immediately accepted. It was not until 1925 that the use of parachutes was formally adopted by the Royal Air Force.

Both Meares and Orde-Lees could therefore be regarded as men of comparable background and attitudes. Both had military experience. Both had served on Antarctic expeditions. Both were known to challenge authority. Orde-Lees' elder brother, a captain in the Border Regiment had been involved in the First Battle of Ypres like Meares, but, unlike him, had been killed. But there any comparability ends.

Meares had seen active service in the First World War whereas Orde-Lees had not. Meares had challenged Scott's attitude to the use of dogs and ponies for transport with some justification. Orde-Lees' appointment as a 'motor expert' to Shackleton's expedition had not been utilised beyond a few unsatisfactory experiments before the expedition sailed from England. Meares had got on well with the other members of Scott's expedition. Orde-Lees had been almost universally distrusted and disliked by the other members of the *Endurance* expedition.

When Shackleton had left twenty two of his men on Elephant Island to go and fetch help in South Georgia, Orde-Lees had been the one man to cause the most difficulty to Frank Wild, the expedition's second in command whom Shackleton had left in charge of the men. Whereas Meares had been regarded as something of an eccentric by his fellow explorers, this was because of his travels and adventures in many different countries, his sense of unusual dress and sometimes unkempt appearance and his apparent cavalier attitude to authority. Orde-Lees was regarded as eccentric because of his pedantic manner and his unwillingness to tolerate the habits and attitudes of his fellow explorers. In fact Orde-Lees was to dislike intensely the habits of the non-commissioned officers and men of Shackleton's party.

There is no record of how Meares and Orde-Lees got on together during the Mission to Japan or whether they got on at all. While Meares left the Mission at the end of November 1921 Orde-Lees stayed on until the end in 1924. He then decided to stay on in Japan and only left – with his Japanese wife – in 1941.

At the beginning of the Mission each of those involved were given honorary ranks in the Japanese Navy. Meares was named as a Commander – the equivalent to a British Army lieutenant-colonel. He must have impressed his Japanese hosts as he was given a farewell dinner by Admiral Kato – the

Japanese Minister of the Navy – at the Maple Club (the Japanese Naval Club) on the eleventh of November 1921.

Then there was a formal banquet in Meares' honour on the seventeenth of November 1921, just before he left Japan. Rear-Admiral Tajiri, commanding officer of the Japanese Training Station at Kasumiguara near Tokyo, where much of the Mission's work had been undertaken, gave a presentation to him and said :

"Commander Meares – I have the honour and pleasure to express to you our deep sense of appreciation of the very valuable services that have been done by you to our Naval Air Service since your arrival to Japan in advance of the British Air Mission.

"As your term of engagement is approaching to a close, you are bound to leave for England before long, but please be assured that you will never be forgotten by us all, and that [it] shall be our unanimous desire to hope that you will safely return to your home and friends after enjoying all the long voyage to the Far West.

"Well, Commander, will you kindly receive a sword which we officers of this Air Station present to you as a small token of our friendship.

"Now, gentlemen, will you all join me in drinking the health of Commander Meares – 'Banzai' ".

The note given to Meares with the sword, which was signed by forty five officers of the Imperial Japanese Navy read : "We undernamed officers belonging to the Kasumigaura Naval Air Training Station have very much pleasure in presenting to Commander C H Meares (attd), I J N, a sword as a small token of friendship on the occasion of the Commander's departure homeward on the expiration of the term of his engagement to the I J N [Imperial Japanese Navy]."

Eight days later Meares received a letter from the Captain attached to the Kasumiguara station :

"My dear Commander, on the eve of your departure for England I hasten to assure you of my very deep appreciation of all you have done for the British Aviation Mission. As you were the first to come out here the foundation which you laid would naturally determine the successful development of our programme, and it becomes more evident each day that the value of your original work is incalculable.

"The success we have to date achieved is chiefly due to your hard work in the early days and your tact in dealing with the many problems that arose. I need not say that the Imperial Japanese Naval Air Service owe you a very great debt and this is fully appreciated by all ranks. The assistance you have given and the unfailing kindness you have shown all Officers and Warrant Officers of the B.A.M will always be remembered.

"Quite frankly, had I originally known that you would have been unable to stay with us for more than a year I would not so readily have considered an extension of our work beyond that period. However, I did not realise it and so we must do our best in spite of the disadvantage caused by your departure. My regret is very real but I know no post I can offer you will make you alter your decision."

Meares was also given the 'Insignia of the Third Class of the Order of the Sacred Treasure in recognition of valuable services rendered by you in connection with aviation in Japan'. He wrote to the British Ambassador in Tokyo for permission to wear the Japanese medal with his War Medals – and the Polar Medal awarded to him as a member of the *Terra Nova* expedition. The British Embassy referred his request to the Foreign Office and, after consulting Buckingham Palace, they gave agreement.

Orde-Lees and senior pilot instructors on the Mission were awarded the Order of the Rising Sun, fourth class at the end of the Mission; many of the others (though, surprisingly not all) were awarded the fifth class of the Order. Orde-Lees was also awarded Japan's Imperial Aviation Society's Medal of Merit.

Opinions in Britain about the Mission varied. There had been some benefits for Britain. Nearly one hundred British manufactured aeroplanes – including single-seaters, seaplanes, flying boats and torpedo carriers – were bought by Japan during 1921 and 1922. But, although there deliberately was not given much publicity to the Mission, there was increasing concern in diplomatic and military circles about its effect and the benefit the Imperial Japanese Navy might reap from it. There was some demand for the Mission to be terminated and the members brought home early. The Foreign Office's response was that "the Mission's work was almost over and the worst of the mischief has been done" [EIB p305].

Perhaps the most succinct comment on the British Aviation Mission was in an article in the Journal of Strategic Studies in September 1982 [JF : A British 'Unofficial' Aviation Mission and Japanese Naval Developments, 1919-1929] : "In 1919 the Royal Navy led the world in naval aviation. It no longer did by 1939. This was partly due to British aid provided at a crucial moment in the development of the Imperial Japanese Navy's air service. Britain assisted Japan to create an air force able to effect the exercise of maritime power.

British decision-makers obviously did not realise quite what they were creating."

VICTORIA B.C.

After his return to England Meares and Christine decided they would travel to various countries and, indeed, they never seemed to stay long in any one place until they finally settled in Victoria, the capital of Vancouver Island in Canada. At one point an English magazine [unknown] in May 1925 featured a photograph of them sitting in chairs at Aix-les-Bains under the by-line 'Water Not Prohibited : Colonel and Mrs Meares engaged in the fashionable pastimes of taking the waters and watching the passing pageant'.

In 1928 the Santa Barbara Daily News of California published a long interview with Meares, saying "Col. And Mrs Meares are spending a few days in Santa Barbara before returning to Victoria, BC, where they are making their home for the present, having searched the world over for a spot in which to retire after many years of active and thrilling service ... Colonel Meares has spent the ten years since the end of the world war looking for a place to make his home. During that time he has visited practically every city in the world as well as hundreds of hamlets and villages and coast towns of every country."

The newspaper then quoted Meares : "Santa Barbara is one of the most charming spots in the whole world. I would make my home here but for one reason. I believe an even climate makes one sluggish. That is why I am returning to Victoria, but I expect to return each winter for two months in Santa Barbara."

The article continued : "Canada has a great future, according to Colonel Meares, who said today that he believes the British immigration office is a little lax in impressing the people of England with the wonderful opportunities which await the colonist in Canada". Meares was then quoted again : "The trouble is that England isn't acquainted with her colonies. When the war broke it is said that some of the leading statesmen of the nation weren't aware whether the colonists in New Zealand are black or white – New Zealand, which is more English than England itself.

"Other countries of Europe see the great future which lies ahead of Canada and thousands of immigrants are pouring into the country while the people of England, who need such an opportunity, remain at home. The farmer class of England, the finest stock of the nation, needs Canada."

Meares still retained his interest in the Antarctic. In 1929 he was one of those who broadcast a message to the Byrd Antarctic expedition from a radio station [KDKA] in Montreal.

In May 1926 Commander Richard E Byrd of the United States Navy, together with his co-pilot Floyd Bennett, claimed to have flown from Spitsbergen to the North Pole and back in fifteen hours. In 1928, then a Rear-Admiral, Byrd led an American Antarctic expedition to the Ice Barrier by the Ross Sea and established a base on the Barrier which he called 'Little America'. As well as exploring the surrounding King Edward VII Land Byrd claimed, in November 1929, to have flown over the South Pole and back. The press reference to the broadcast at the end of March 1929 contained a message to Meares from Byrd : "we all enjoyed the special Scott program from KDKA and your message wishing us many happy returns of the daylight. It is a pleasure to receive the greetings of veterans of the Antarctic like you and Admiral [Teddy] Evans and we send you cordial good wishes from Little America. Byrd"

THE ANTARCTIC CLUB

In January 1929 The Antarctic Club was formed with Reginald Skelton, the physicist with Scott's *Discovery* expedition of 1901 to 1904, as its President. Membership of the Club was "restricted to members of British Expeditions which have been engaged in exploration work in the Antarctic Regions". Meares was admitted to membership on the thirteenth of June 1929 and was given certificate number twenty eight.

Another member of The Antarctic Club, Louis Bernacchi , the Belgium born Australian who had been a physicist with Scott's *Discovery* expedition , called on him in Santa Barbara in1935. In the Times obituary of Meares in May 1937 Bernacchi had written "I have a happy memory of spending a day with Meares and his charming wife at his Santa Barbara bungalow on the Californian coast less then two years ago, and of a delightful drive in his car along 100 miles of picturesque sea coast to Hollywood".

In the years after the death of Captain Scott there were many books and biographies of the members of his expedition. Scott's wife, Kathleen, had been consulted about Scott's diaries by the leading members of the Royal Geographical Society and they were published under the title 'Scott's Last Expedition' in two volumes by Smith & Elder of London. They were well received. Bernacchi had written the first biography of Captain Lawrence Oates – 'A Very Gallant Gentleman'. It was published in 1933 and Bernacchi gave Meares a copy with the signed inscription ' With compliments and thanks, April 1933'.

Meares and his wife liked Victoria and over the years lived in different houses there. Meares always maintained that he was a descendant of the former fur trader and lieutenant in the Royal Navy, John Meares, after whom a street in Victoria and a small island off the coast, were named.

John Meares resigned from the Royal Navy to make his fortune as a fur trader. China was then – in the latter part of the eighteenth century – a good market for furs. In 1786 John Meares had sailed, in a ship called *Nootka*, from China to Alaska. He was one of the main people to develop the port of Nootka which then became the busiest port in the north west coast of the North American continent. There were difficulties when Spanish ships tried to claim the area round Nootka for the Spanish Crown. The British Admiralty sent out Captain George Vancouver of the Royal Navy to wrest the area away from Spain and secure it for the British Crown. The Daily Colonist newspaper of Victoria, in an article in June 1939, claimed that "It was largely upon the discoveries and reports of the two men, Meares and Vancouver, that Great Britain was able to substantiate her claim to the coastline of the territory now known as the Province of British Columbia".

There is no record or evidence of a family connection between the eighteenth century John Meares and the twentieth century Cecil Meares, but it would be nice to think there was such a connection.

Meares died in the Jubilee Hospital in Victoria, after a short illness, on the twelfth of May 1937. His death certificate put the cause of death as 'Cholaemia due to hepatic cirrhosis and cholecystitis'. He was sixty years old. His wife Christine outlived him until 1974 when she died in a Victoria nursing home, at the age of eighty one.

Meares and Christine had no children of their own. Meares, in his will, had left all his possessions and assets to Christine. In her will she left her husband's medals, Japanese sword, articles and letters to the Royal British Columbia Museum. Her other possessions she left to be divided among eight different people – three nieces, two cousins, a sister-in-law and two friends. Attempts at contacting any relatives of these have been abortive. Meares himself also decreed in his will that on his death he should be cremated and that his ashes "be sprinkled in some convenient woods or garden". It was a modest departure from life – but a life of extraordinary adventure.

Meares never sought publicity for himself. He did write several articles for magazines on his travels in the Far East and in South-East Asia, but he never wrote anything about his time with the *Terra Nova* expedition nor during the First World War.

In many of the books about Scott's last expedition there has been debate about whether Scott's party could have survived. They had got to only eleven miles from the supplies at One Ton Depot. On the other hand that would still have placed them many miles from their base at Cape Evans and, whatever nourishment they would have been able to get from those supplies, Scott's feet were badly frostbitten and it is doubtful whether he could have gone much further. Could Wilson and Bowers have gone on without him to fetch help? They were weak and exhausted and it would have been over one hundred miles to go for help.

The debate also poses questions about whether the dogs could have been used to rescue Scott's party. Should Cherry-Garrard and Dimitri have gone further south than One Ton Depot to look for Scott? The weather was appalling – and Scott, Wilson and Bowers, had been held by a blizzard for nine days – but would it have been possible? The answer, of course, is that no one can say with any certainty. Could Cherry-Garrard and Dimitri have possibly found the Scott party? It was a question that was to haunt Cherry-Garrard for the rest of his life.

If Meares had stayed with the expedition for the second winter – and not gone back to New Zealand with the *Terra Nova* before the second winter set in – would that have made any difference? Meares would have been far better with the dogs than Cherry-Garrard and he and Dimitri might have gone beyond One Ton Depot to look for the Scott party but the weather would still have been awful, Scott's party would still have been held up by the blizzard and in such weather it would still have been very difficult to find them.

But Meares had made it clear that he was only staying with the expedition for the one winter. He had also – and crucially – assisted in the Southern Party's journey carrying food and equipment across the Barrier and up on to the Beardmore Glacier. If Meares had turned for home at 81° 30' south as Scott had originally intended, then the progress of Scott's party would have been slower and would they have got back to as near to One Ton Depot as they did do?

Meares had played a significant part in the depot-laying party in the Antarctic summer of 1911 and also in his return from the Beardmore Glacier in December 1911. That journey – though extremely difficult for him and Dimitri, and on short rations, – had been useful for the return of Scott's supporting parties in clearing the food depots laid on the way out.

The wider question of whether Scott could have got to the South Pole and back – let alone beaten the Norwegian, Roald Amundsen, to the Pole – if

Meares and the dogs had gone with him all the way is something it is of course impossible to answer : as are the questions of whether Scott would have been in a better position by only taking three other men with him to the Pole instead of four others. And there is the whole issue of Scott's preference for ponies as opposed to dogs and the traditional British reliance on man-hauling.

Whatever the questions, which naturally remain unanswered, and whatever the speculations, the fact remains that Meares acquitted himself well and was a credit to Scott's expedition. Prior to the expedition he had led an extraordinary life of travel and, after the expedition, had served with distinction in the First World War. His life was varied in the extreme but, especially because of his reluctance to court publicity, perhaps he will always be known simply as 'Scott's dog driver'.

Whatever the disagreements between Meares and Scott it is worthy of note that in 1913 Meares had been given a copy of the two volume book 'Scott's Last Expedition'. Inside was the inscription : "Cecil Meares, With best wishes from Kathleen Scott".

REFERENCE NOTES :

AFRT	A First Rate Tragedy – Captain Scott's Antarctic Expeditions By Diane Preston
AT	Alexander Turnbull Library, Wellington, NZ
BC	British Columbia Archives, Royal BC Museum, Canada
Deb	The Quiet Land – the Antarctic Diaries of Frank Debenham
EIB	Elephant Island & Beyond by John Thomson
EW	Edward Wilson – Diary of the Terra Nova Expedition to the Antarctic 1910 - 1912
FS	Captain Scott by Ranulph Fiennes
JF	John Ferris, Journal of Strategic Studies, September 1982
L & C	Captain Oates – Soldier and Explorer by Sue Limb and Patrick Cordingley
LPE	The Last Place on Earth by Roland Huntford
OMRS	Orders and Medals Research Society – Journal, Summer 1997, article by Wing Commander Jim Routledge
RAF	Royal Air Force Museum, Hendon
RFC RB	The Royal Flying Corps in World War One by Ralph Barker,
SofA	Scott of the Antarctic by Elspeth Huxley
S	Scott's Last Expedition
Si	Silas – The Antarctic Diaries & Memoir of Charles S Wright,
SPRI	Scott Polar Research Institute, Cambridge
SwS	South with Scott by Captain Edward R G R Evans
TBP	The Blue Peter, volume XII June 1932
TGW	The Great White South : or with Scott in the Antarctic 'by Herbert G Ponting
WJW	The Worst Journey in the World by Apsley Cherry-Garrard, 2 volumes
WNF	Adventure, Sport & Travel on the Tibetan Steppes by WN Ferguson

TERRA NOVA 1910 - 1913

Officers :
Robert Falcon Scott	Captain RN
Edward Evans	Lieutenant RN*
Victor Campbell	Lieutenant RN
Henry Bowers	Lieutenant RIM
Lawrence Oates	Captain Iniskilling Dragoons
Murray Levick	Surgeon RN
Edward Atkinson	Surgeon RN, parasitologist

Scientific Staff :
Edward Wilson	Chief of scientists, zoologist
George Simpson	Meteorologist*
Griffith Taylor	Geologist*
Edward Nelson	Biologist
Frank Debenham	Geologist
Charles Wright	Physicist
Raymond Priestley	Geologist
Herbert Ponting	Camera artist*
Cecil Meares	In charge of dogs*
Bernard Day	Motor engineer*
Apsley Cherry-Garrard	Assistant zoologist

Men :
William Lashly	Chief stoker RN
WW Archer	Steward, late RN**
Thomas Clissold	Steward, late RN*
Edgar Evans	Petty officer RN
Robert Forde	Petty officer RN*
Thomas Crean	Petty officer RN
Thomas Williamson	Petty officer RN**
Patrick Keohane	Petty officer RN
George Abbott	Petty officer RN
Frank Browning	Petty officer RN
Harry Dickason	Able seaman RN
Frederick Hooper	Steward, late RN
Anton Omelchenko	Groom*
Dimitri Gerov	Dog driver

* denotes those who returned on the *Terra Nova* in March 1912
** denotes those who arrived on the *Terra Nova* in March 1912

BIBLIOGRAPHY

A First Rate Tragedy – Captain Scott's Antarctic Expeditions by Diane Preston, Constable 1997

Adventure, Sport & Travel on the Tibetan Steppes by WN Ferguson, Charles Scribner 1911

Antarctic Adventure – Scott's Northern Party by Raymond E Priestley, T Fisher Unwin 1914

Antarctic Days by James Murray and George Marston, Andrew Melrose 1913

Bartlett – The Great Canadian Explorer by Harold Horwood, Doubleday 1977

British Polar Exploration and Research – A Historical and Medallic Record with Biographies 1818 - 1999 by Lieutenant Colonel Neville W Poulsom & Rear Admiral JAL Myers, Savannah Publications 2000

Captain Oates – Soldier and Explorer by Sue Limb and Patrick Cordingley, BT Batsford 1982

Captain Scott and the Antarctic Tragedy by Peter Brent, Weidenfeld Nicolson 1974

Captain Scott by Ranulph Fiennes, Hodder & Stoughton 2003

Cherry – A Life of Apsley Cherry-Garrard by Sara Wheeler, Johnathan Cape 2001

Cook & Peary – The Polar Controversy Resolved by Robert M Bryce, Stackpole Books 1997

Elephant Island & Beyond – the Life and Diaries of Thomas Orde Lees by John Thomson, Bluntisham Books 2003

Edward Wilson – Diary of the *Terra Nova* Expedition to the Antarctic 1910 - 1912 edited by HGR King, Blandford Press 1972

Farthest North – The Norwegian Polar Expedition 1893 - 1896 by Fridtjof Nansen, 2 volumes, Archibald Constable and Company 1897

Frank Wild by Leif Mills, Caedmon of Whitby 1999

High Latitude by JK Davis, Melbourne University Press 1962

Icebound – The *Jeanette* Expedition's Quest for the North Pole by Leonard F Guttridge, Airlife 1988

Karluk – the great untold story of Arctic exploration by William Laird McKinlay, St Martin's Press 1976

Mind Over Matter by Ranulph Fiennes, Sinclair-Stevenson 1993

My Attainment of the Pole – being the record of the expedition that first reached the boreal centre with the final summary of the polar controversy by Dr Frederick A Cook, Polar Publishing Co. 1911

Photographer of the World – a biography of Herbert Ponting, by HJP Arnold. Hutchinson 1969

Pilgrims on the Ice – Robert Falcon Scott's First Antarctic Expedition, by T. H. Baughman University of Nebraska Press 1999

Scott of the Antarctic by Elspeth Huxley, Atheneum 1978
Scott's Last Expedition, Smith & Elder 2 volumes, 1913
Shackleton by Margery & James Fisher, James Barrie 1957
Shackleton by Roland Huntford, Hodder & Stoughton 1985
Silas – the Antarctic Diaries & Memoir of Charles S Wright edited by Colin Bull & Pat Wright, Ohio State University Press 1993
South with Scott by Captain Edward RGR Evans, W Collins 1923

The Friendly Arctic by Vilhjalmur Stefansson, George G Harrap 1921
The Great White South : or with Scott in the Antarctic by Herbert G Ponting, Gerald Duckworth 1950

The Heart of the Antarctic – being the story of the British Antarctic Expedition 1907 - 1909 by EH Shackleton, 2 volumes, William Heinemann 1909
The Ice Master by Jennifer Niven, Macmillan 2000
The Last Place on Earth by Roland Huntford, Hodder & Stoughton 1979
The Log of Bob Bartlett by Captain Robert A Bartlett, Blue Ribbon Books 1928
The North Pole – with an introduction by Theodore Roosevelt – by Robert E Peary, Hodder & Stoughton 1910
The North West Passage – the Gjoa Expedition 1903 - 1907 by Roald Amundsen, Archibald Constable and company 1908
The Quiet Land – the Antarctic Diaries of Frank Debenham edited by June Debenham Back. Bluntisham Books 1992
The Royal Flying Corps in World War One by Ralph Barker, Robinson 2002
The Search for the North West Passage by Ann Savours Shirley, Chatham Publishing 1999
The South Pole by Roald Amundsen, John Murray 1912
Stefansson and the Canadian Arctic by Richard Diubaldo, McGill-Queen's University Press 1978
The Voyage of the *Discovery* by Captain Robert F Scott, 2 volumes, John Murray 1905
The Worst Journey in the World by Apsley Cherry-Garrard, 2 volumes, Constable 1923
William Speirs Bruce – polar explorer and Scottish nationalist by Peter Speak, National Museums of Scotland Publishing 2002

GLOSSARY

Aneroid : aneroid barometer measures atmospheric pressure without the use of fluids; it consists of a metal (tin) box, partially exhausted of air, where the thin lid expands or contracts according to changes in air pressure

Bergschrund : in a typical glacier the upper portion is hidden by neve and often by freshly fallen snow and is smooth and unbroken; during the summer, when little snow falls, the body of the glacier moves away from the snow field and a very deep crevasse is usually formed – a bergschrund

Brash ice : small fragments of broken ice, the wreckage of other forms of ice

Crevasse : a deep fissure in ice

Fast ice : unbroken sea ice of varying width which remains attached to the coast

Finnesko : boots made of reindeer skin, suitable for snowy conditions but offering little grip on ice

Floes : flat frozen surface of the sea – a free piece of sea ice; ice up to three feet thick is usually described as 'light floes' and thicker floes are 'heavy floes'; floes can be ten miles or more in diameter

Gneiss : laminated rock of quartz feldspar and mica

Hoosh : a thick soup usually made by melting snow with pemmican and, sometimes, plasmon biscuit

Hummocked ice : sea ice piled haphazardly one piece over another

Hypsometer : an instrument for measuring altitudes by determining the boiling point of water at a given altitude

Ice sheet : a continuous mass of very thick ice extending many miles

Ice shelf : a floating shelf of thick ice; the seaward cliffs of ice shelves range from six to two hundred feet in height

Ice wall : an ice cliff forming the seaward margin of an inland ice sheet

Neve : compacted snow – halfway between soft loose snow and glacier ice, old snow which has been compacted into denser material

Nunatak : an island like outcrop of rock projecting through a sheet of surrounding land ice; a rock mass, often shaped like a pyramid, entirely surrounded by ice and snow; usually the top of a buried hill or mountain

Pack ice : loose ice from broken ice floes which might also contain pieces of icebergs; any area of sea ice other than fast ice

Pemmican : dried and pounded meat (usually beef) mixed with fat

Pressure ice : a general term for sea ice which has been squeezed and in places forced upwards – then described as hummocked ice or pressure ridge

Sastrugi : rough surface of snow caused by continuous winds blowing across the snow surface – also called snow waves – varying in size according to the force of the wind and the compactness of the snow; fluted ridges carved by the wind from a snow surface – from a few inches to several feet high

Water sky : dark patches in the sky due to the reflection of open water on clouds

Whiteout : a condition of diffuse light when no shadows are cast, due to a continuous white cloud layer appearing to merge with the white snow surface; no surface irregularities are visible – there is no visible horizon

* Temperatures given in the text are all in fahrenheit
* Miles quoted are geographical/nautical miles – one mile corresponding to one minute of latitude and therefore sixty miles equalling one degree of latitude – that is 6,082 feet to the mile as distinct from a statute mile of 5,280 feet

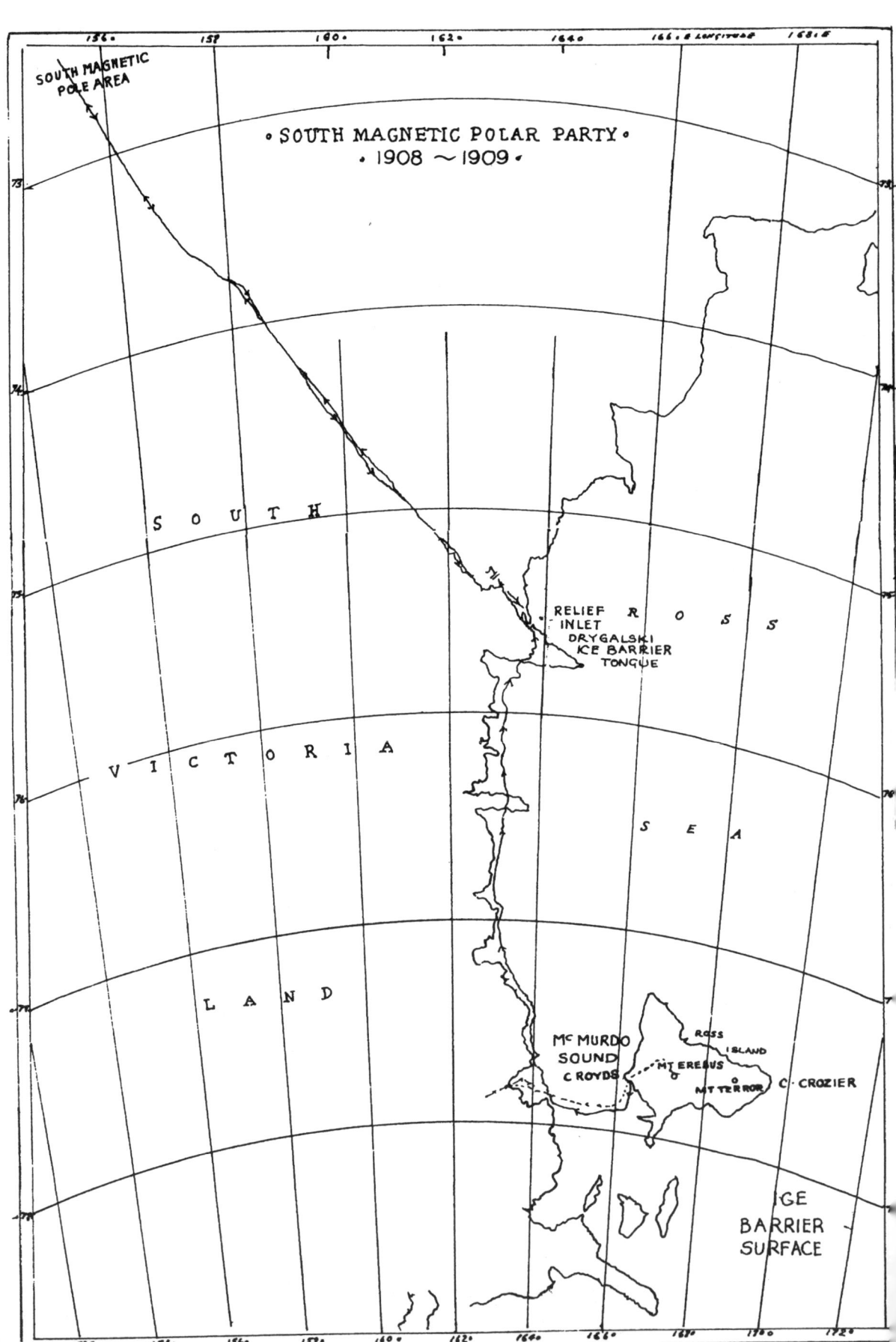

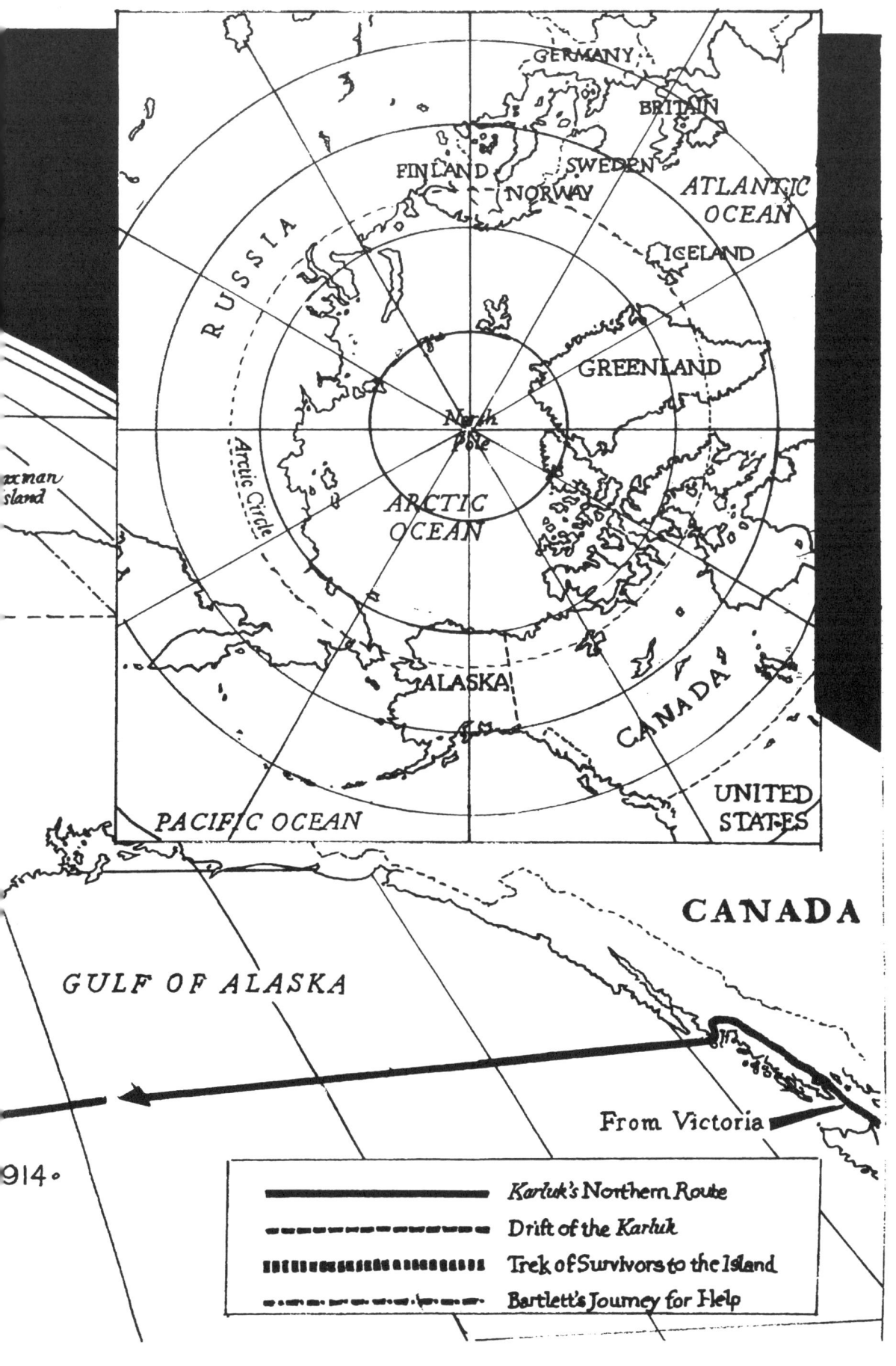